Praise for
The Last Wilderness

"I am impressed by how Michael has been able to combine business development with concepts of environmental protection in his work. It is incumbent
on each one of us think about the environment in our day-to-day lives and
what things we can do—big or small—to make the world a better place to
live for our children. Michael's efforts in Kachemak Bay demonstrate this
principle."

—**Peter Gilruth**, director of the
Division of Early Warning and Assessment,
United Nations Environment Programme

"A captivating account of decades in the Alaska wilderness as Michael and
Diana McBride literally respond to the 'call of the wild.'... Destined to be not
just an Alaskan classic, but one of nature and wilderness writ large. A must
for any nature bookshelf."

—**Tom Lovejoy**,
founder of the PBS series *Nature*

"Michael McBride is a thinker, but he's also a doer, and what he's done
could fill a book: this one. I've stayed at the Kachemak Bay Wilderness
Lodge, spent time with Michael, enjoyed his wisdom, his stories, his
hospitality. These pages brought me right back to his side, right back to the
fireside, right back to China Poot Bay, right back to paradise. Beautifully
written, generous in spirit, even magnanimous, *The Last Wilderness* is a
richly satisfying read. And it might just send you on your way to Alaska."

—**Bill Roorbach**, author of *Temple Stream*,
Summers with Juliet, and *Life among Giants*

"Southern Alaska has a strong claim to the title of 'Best Remaining Place in the World.' And few have known it as intimately, as well—or as lovingly—as Michael McBride. Read his story, but don't just read it—go out and ground-truth what he says and see for yourself what he means. Alaska is one of the few places that really needs you. And we need it. McBride's narrative explains why, and in doing so, it succeeds wonderfully."

—**Carl Safina**, author of
Song for the Blue Ocean

"Michael McBride gave his heart to Alaska half a century ago, and he fulfilled his dream in a remote spot of wilderness. Like an old-time pioneer, he was adept at survival, building his own lodge, becoming a bush pilot, raising a family. I read of his joyful struggles with greatest admiration as with compassion he describes his life and that of the local peoples and the wild animals. And he never relents in his fight to protect Alaska's beauty."

—**George Schaller**, author of *Tibet Wild*

"Michael McBride is the perfect person to chronicle this creative and loving relationship between human stewardship at its educational best and the beauty of wilderness at its most extraordinarily profound."

—**Vernon D. Swaback**, FAIA, FAICP
author of *The Creative Community*

"Michael's book rekindles memories of wilderness adventures for those who have experienced wild Alaska. For those who have not yet known the thrill of a close encounter with a brown bear or harbor seal, it is a beautiful invitation to experience Alaska's best."

—**Fran Ulmer**
former lieutenant governor of Alaska

"McBride's book is a rhapsodic symphony celebrating his half century odyssey to create an idyllic family life in the Alaskan wilderness. There he began with virtually nothing and built a 'something' that is now a model of ecotourism. Open this work at any place and you are treated with a philosophy of the natural life that would be the envy of a John Burroughs or John Muir."

—**Captain Don Walsh**, US Navy (retired), PhD
oceanographer and explorer

"The charisma and energy that made Michael McBride an Alaskan outdoor legend flow through these pages like a school of salmon charging upstream, powered by the love and ecstatic joy that link him to his home shores of Kachemak Bay. If John Muir were alive today, this is how he would live and this is how he would write."

—**Charles Wohlforth**,
author of *The Fate of Nature*

"This book is a tribute to life, family, and a man's love and commitment to preserving the wilderness not only in Alaska, but in the world. It is an eloquent plea to all of us to recognize the need for saving the wilderness. But, the book is also a personal adventure story of an incredible man and his incredible family."

—**David D. Goodenough**, founder of
Goodenough Company, Inc.

The LAST
WILDERNESS

The LAST WILDERNESS

Alaska's Rugged Coast

MICHAEL MCBRIDE

Fulcrum Publishing
Golden, Colorado

Library of Congress Cataloging-in-Publication Data

McBride, Michael, 1943–
   The last wilderness / Michael McBride.
      pages cm
   Includes bibliographical references.
   ISBN 978-1-938486-37-1
   1. McBride, Michael, 1943- 2. Kachemak Bay Region (Alaska)–Biography. 3. Kachemak Bay Region (Alaska)–Social life and customs. 4. Frontier and pioneer life–Alaska–Kachemak Bay Region. 5. Outdoor life–Alaska–Kachemak Bay Region. 6. McBride family. I. Title.
   F912.K26M35 2013
   979.8'3–dc23
   [B]

2013017166

Printed in the United States of America
0 9 8 7 6 5 4 3 2 1

Fulcrum Publishing
4690 Table Mountain Drive, Suite 100
Golden, CO 80403
800-992-2908 • 303-277-1623
www.fulcrumbooks.com

Contents

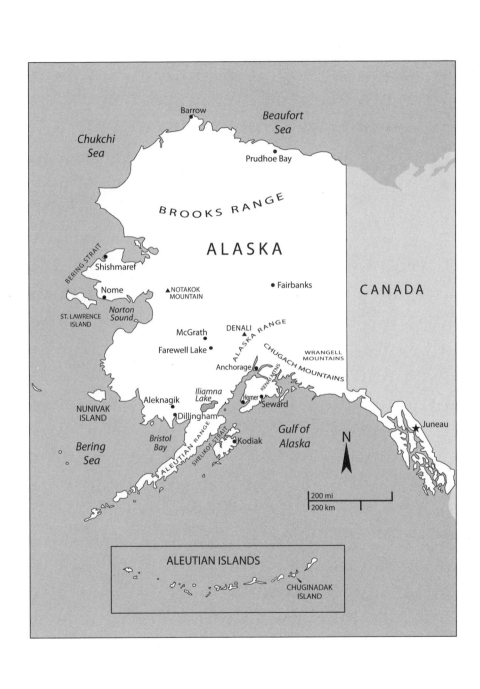

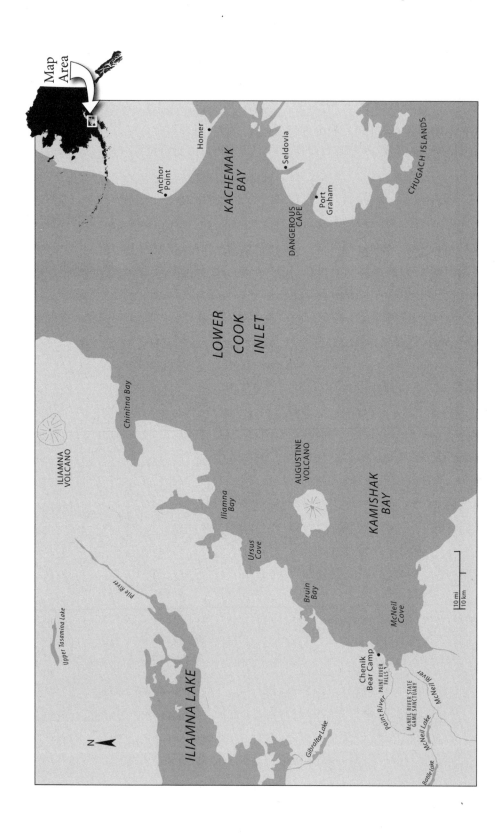

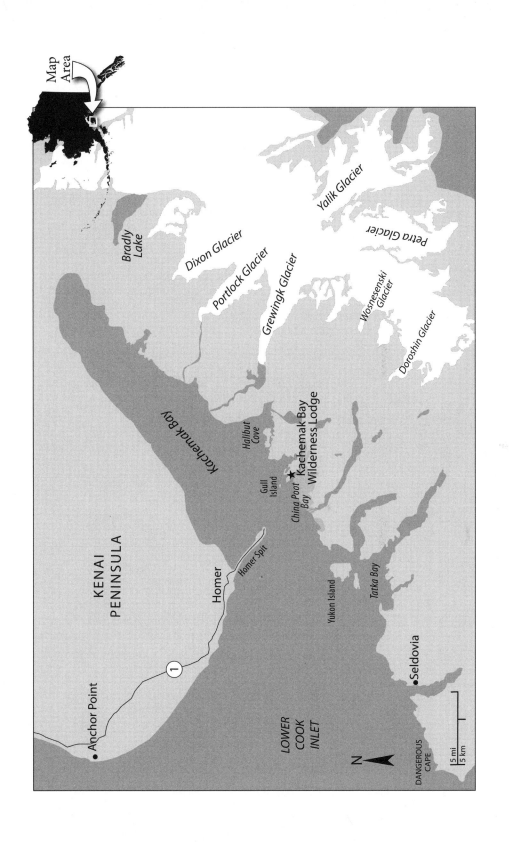

Anangula,
Whale Swimming North

These stories began to take shape atop a three-legged tripod of Alaskan experiences. The ancients said, "Chuginadak breathes through a woman," referring to an active volcano in the uninhabited Islands of the Four Mountains in the Aleutians. There was once a village that the people called Anangula, or "Whale Swimming North," on Umnak Island within sight of Chulaka. On the Bering Sea coast northwest of Togiak in far west Alaska, the Eskimo warrior Apanvugpak rests in eternal sleep. These little-known places are each the catalyst for fabulous stories of places so remote and unknown that few Alaskans have ever even heard of them. There are myths and legends and thousands of years of human habitation whose stories are as valuable and informative as any on earth. They serve to remind us that even when we think that the world has become very small, we find there are marvelous discoveries to be made right on our home ground.

The window through which we saw Alaska as a young couple in the 1960s was fast closing. By and large people today have left the remote places and returned to electricity, accessible school systems, emergency services, and Internet access. I am pained by this in a way I don't quite understand. It seems that the land is lonely without the people who, more than anything, loved being in wild places.

Getting to know our fellow Alaskans was going to change our lives forever. Associations with the two-leggeds and four-leggeds would invigorate us, embolden us, intimidate and remind us that we were alive—completely alive. A sense of wonder would descend on us as we tried to see the world through the salmon's eye, smell what was within the woodsmoke and cottonwood buds, hear other voices in the loon's cry and the wind's whisper, and taste anew, each time we drank the water from the river.

We have been asked repeatedly to write down not only these stories but also the story of how we came to the roadless place we call home just a few years after Alaska's transition from territorial status to statehood. People wanted to know how we first survived and then prospered beside our northern estuary. Many are those who went to more distant places and lived rich and fulfilling lives—there are many fish and bigger fish in this pond, and ours is but a minnow. We are, each of us, much more the product of the times than producers of the situations around us.

Each of us is on a remarkable journey; each of us has a story worth telling, worth saving. Our own little adventure is simply a retelling of a story that has been told a thousand times in countless places the world over. We saw the world through our own prism of past experiences, which in turn shaped our vision of the future—what was possible—and propelled us forward.

What I have set out to do is repay an overdue debt. I want to sing a praise song to honor the source that allowed us to have the rich tapestry of experiences that wrapped around us like a red-and-black striped Hudson's Bay three-beaver blanket. I want very much to sing it well, enunciating each syllable carefully in the way that ancient songs and stories were always repeated. I want it to bring smiles, tears, and cause the nodding of heads, indicating resonance with the listeners that they too have felt what we felt and that there is camaraderie among the community of humans and among kindred spirits.

What we did in the late 1960s was nothing so special; we did what others were doing all across Alaska. The fact is, however, that today, very few people are out there living on the land as we were, prepared for isolation and long, lonely dark winters and ready to go like hell during the brief summers. During our short-lived era, the state actually sent schoolteachers out from Juneau to ride circuit to homesteads and remote cabins to check in with their correspondence students once a year. Once there, they often spent days sleeping on the floor or in a shed—no one out there had an empty spare bedroom. There was a need for this kind of program, and the state supplied it. It has since disappeared with those scattered families who have moved closer to medical services, to the radius of cell phone coverage, and to be with other people. Chances are that any of them who now have TV would probably agree that they would be better off without that damned intruder in their home. Bob Dylan sang about all of this as it was a'hap'nin, and it was true in the Alaska bush as well: The times they were a-changin'.

We were among the luckiest of our group, for we found a little niche that could be amplified to accommodate us throughout a lifetime. We are optimistic that our children, grandchildren, and subsequent generations will be able to have a somewhat similar experience in the same place. Instead of being a part of the curling wave that rolls across a continent, perhaps the future generations will live on a shoreline that is more static in nature, without neighbors constantly moving west, and that will in turn offer its own challenges and rewards. Perhaps their explorations will be of a sort we can't imagine; perhaps theirs will be under the sea, into space, or on journeys deeper into their own hearts and minds.

Unlike the people of only a few generations ago, we have come to the realization that the way we are living on this planet is not sustainable, and that we are in immediate danger of fouling the very nest that incubated us.

A sense of wonder inspired by the natural world and the beautiful people in it is, I believe, at the center of the motivation to act. We experience, we learn, we act—the three-legged tripod of sustainability is a beacon to follow as we plot a course for the future.

Acknowledgments

The unsung hero in this story is my long suffering wife who, through-out it all, lipstick in hand, remained "pretty as a rose and tough as a boot." The praises I sang for her in the text are woefully inadequate. Thank you, dear! Daughter Shannon and son Morgan were also guides, counselors, and mentors, and I am deeply indebted to them. Author/lifelong sidekick, M'zee Boyd Norton, like the woodpecker, kept knocking at my door reminding me to write the book, and editor/publisher/friend Bob Baron, with prodding and praise, artfully turned plan into reality. It is these good people and my expanded circle of friends, my tribe, whom I honor with "the praise of the current for the source." My praise song was written to celebrate all who are working to make this blue and green world a better place. May there be blessings upon you, my dear beautiful friends. May we go forward honoring the land and the sea and the sky, each other, our teachers, our ancestors while walking in a sacred way.

The Homer Spit. Photo © Boyd Norton.

CHAPTER 1

Crossing the Bay

"All journeys have secret destinations of which the
traveler is unaware."

—*Martin Buber*

The Caterpillar diesel beat in a steady drone beneath the floorboards of the old barge. Its throbbing vibrated up through our cannery boots and rattled the plates in the galley. It was reassuring, sonorous. It quieted the nervous anxiety we felt crossing the bay that was lashed with rolling, sawtoothed whitecaps. Although she didn't say so, I knew Diane was scared. The view from the wheelhouse windows revealed a frightening scene. Alex Flyum, the old Bering Sea Eskimo who owned the boat, stared intently ahead; he was a fellow of few words. Diane and I, with Alex's little wife, Annie, braced ourselves solidly as the rusty old World War II landing craft named *Nanuk*, or "Ice Bear," labored sickeningly in the heavy seas.

Back in 1969 — all those years ago — we watched as the little fishing village and boat harbor disappeared as we motored east toward the uncertainties of a new life, we felt like we were leaving for the dark side of the moon. In 1966, the 225 miles of dirt road south of Anchorage, Alaska, to Homer was at times impassable, washed out, snow drifted, or so potholed that you could walk faster than you could drive. Homer and Kachemak Bay were a hundred miles as the gull flies north of Kodiak across one of the toughest ocean entrances anywhere. From that entrance one could draw a straight line to Antarctica and not encounter land.

New to Alaska and a bachelor, when I first saw it, I fell hopelessly in love. I recall it now as if it were yesterday. Standing on the bluff overlooking the bay, I pointed to the prettiest place I could see across the miles of blue-green water and thought to myself, "I'm going to spend the rest of my life right down there."

When Diane and I first arrived, Homer was a young town of about eighteen hundred souls, and its outlying roads were best suited for farm tractors. There was a sleepy harbor and hillsides of birch and spruce that carpeted down from the broad plateau of the Kenai Peninsula to salt water. The ten-by-thirty-mile estuary was protected by a long sandbar that nearly occluded its opening to the sea. The whole bay offered a protected anchorage close to some of the richest fishing grounds in the world. Its equal could be seen only in a few places in all of Alaska. Since it lay at the tip of the westernmost road in America, it was literally the end of the road. The richness of the metaphor took many forms; it was as far as you could go without a passport. There was a common abundance of well-being. I saw a vibrant community with a soul of its own, set apart from the mainstream of humanity and less affected by the bumps and starts of society than other towns. Life had made of it a backwater where everything was quiet, poetic, and deep. Its embracing view of ocean and snowcapped peaks, glaciers and ancient forests comprised what simply had to be one of the most spectacular panoramas in the world. There was a tall snow-covered range of mountains running northeast to southwest that was bejeweled with more than a dozen glaciers, many of them visible from town. On sunny days when the sun hit them at an optimum angle the blue-green luminescence of their terminal faces appeared to actually glow. Those who lived and worked in the midst of their grand visual richness were not jaded by their good fortune; rather, it manifested itself in their characters.

The area was populated by fishermen and homesteaders, artists and poets, ranchers and bush pilots. They were a warm, independent, resourceful people, a few of whom we had come to know and admire in the days we spent in town gathering the last of our winter supplies. The people had an appreciation for and sensitivity to the mood of the land and sea. They were in harmony with the rhythms of the earth, with the sea and sky, just as the ancient residents had been for thousands of years before them. If there was a definable industry, it was tenacity. These people were more closely tuned to their tide tables and almanacs than to clocks and calendars. Theirs was an attachment to lasting values; self-restraint came easily. If a bird's feather fluttered down to the street, a passing resident could pick it up and humorously observe what kind bird was now less flight worthy.

At the post office or grocery store the most commonly exchanged information in the fall was who had recently taken a moose, where, and with whom it had been shared. Hundreds of pounds of beautiful red meat made the difference between a bountiful winter and a lean one. A flock of pigeons lived in one of King's old cabins just off Pioneer Avenue, where they hid from the abundant hawks, owls, falcons, and eagles. Brown bear tracks were occasionally seen on the banks of Fritz Creek, and moose were abundant in Homer once the hunting season closed.

From the intensity of the summer fish camps or work in distant canneries across Cook Inlet came a sense of adventure and willingness to change environments to better their lot. Mingling with seasonal workers and visitors from diverse walks of life made them congenial and broad minded. These were people who would roll out forbearance in the face of tribulation and courage to meet adversity as they had just a few years before, in 1964, when the great earthquake nearly destroyed Kodiak and Seward.

Accustomed to busy and arduous summers when the sun shone almost around the clock, fall offered a time for establishing a perspective on life. A contemplative backward look hedged the bets on future success. Their winters were confining and slow, but the long darkness was made less oppressive by the promise and optimism of the distant spring. The quiet time offered endless opportunities to socialize. Rivers, streams, and muskeg froze, allowing easier travel. Visitors stayed for days. On snowshoes, foot, and skis, by dogsled, horse, and wagon, they converged on gathering places for old-fashioned fun. So limited was public space that the Waterfront and the Club Bar served as convivial meeting spaces for birthdays and other parties. When Benny Bauers played the accordion at the Homer Women's Club, everyone danced long into the night.

It was said that in the days of the gold rush, those who had not found enough color in the creeks to go "outside" for the winter would find their way to Homer. As long as they had enough strength to stagger down to the edge of the water, they would be well fed. Old-timer

Hugh Watson once told me, "A gun and some cartridges, a clam shovel, and a little cabin to saiwash into, and hell, a feller could winter right well."

As we were preparing for our first winter across the bay, we shared many cups of coffee with homemade bread and jam, and even a taste or two of some fine home brew with our new friends while waiting out a long storm. These people had rare combinations of interests and skills. They possessed a sense of place and sense of wonder both for their surroundings and one another. Their view of themselves and the world around them was endearing. Leaving good people like this and the little town behind, we felt tremendously lonely, as though we were making a one-way trip. We didn't know when or how we were coming back. We had no money, having spent our last dollars on supplies. There was nothing to fall back on. We had no boat or means of getting one. We were being tugged and pulled like metal filings by a magnet, knowing that somehow, we could, we would, build a life beside the magical estuary to which we were headed.

Entering this country without a boat was similar to a cowboy walking into the hills without a horse. Our meager pile of possessions — virtually all we owned in the world — was piled below on the deck of the *Nanuk* and lashed down under a ragged canvas tarp. Even though we would be separated from our new friends by a dangerous body of water, the kinship that had blossomed as we prepared for our new life promised rich and eventful relationships in the years ahead. This buoyed our spirits as perhaps nothing else could have.

The stinging north wind reminded us that a severe winter was at hand. If there were things we needed for the hardest part of the year ahead and they were not under that tarp, we weren't going to have them and that's all there was to that. I longed for more traps, a better gun, more ammunition. We had been like kids in a candy store at George Bishop's one-room Trading Post, wanting much more than we had money to pay for. I wondered if we had remembered extra blades for the Swede saw — yes, I thought. And thank goodness, I had remembered to buy the spare Aladdin lamp mantles.

As I stared through the porthole at the storm, it occurred to me that I might actually be guilty and irresponsible in this foolishness. We were focused on a goal that we didn't quite understand. We weren't running from anything; rather, we were in search of something — something intangible that we couldn't quite describe. The sheer power of the adventure unfolding before us must have been propelled by the genes from our ancestors whom we knew had done similar things. Mysteriously we had already fallen under the spell of that place across the bay, and now were responding to its tugging, its gravitational pull, its expanding grip. We had felt it at our first visit, and it had precipitated the crossing of this uncertain watery threshold. The consummation was at hand.

In the decades to come, the commitment would mushroom full blown and limpet-like; we would develop such a fierce tenacity that we would be able to hold on against terrific obstacles. As a student, I had been enamored by the writings of Robert Browning and his description of a love that "reached for the ends of being and ideal grace." I just knew somehow that this was the place, now was the time, and we were the ones who, like the Inupiat word for snowy owl, would be "the ones who stay all winter."

Looking back now, I cannot in my wildest dreams imagine how that pretty little blonde pigtailed girl at my side could have trusted me enough to be a partner in this crazy scheme. But we were in love, and its power was no less than that which drove the blunt bow of the old landing craft through the storm. Looking her way, I saw the strength and confidence in those deep-blue sparkling eyes and it told me all I needed to know. We were centered on our still-new relationship and had an unexpressed confidence that everything else would fall into place. Like Alaska itself, we were fresh and full of spirit. We were frightened a little — perhaps because of the unfamiliarity of our sur- roundings and the uncertainties of our destination. The storm wasn't helping. But it was baptismally fitting, and we accepted it as necessary to our growth. Our compassion, patience, and understanding of one another were introducing us to a life song of calm and storm, wind and stars. The home we envisioned would be the cradle of our dreams and nurture a much stronger love yet to evolve.

Although the events leading up to this point were deliberate, what would come next was tenuous; the unpeopled bay ahead held only challenges. The nearest neighbors were miles away across danger- ous, unforgiving water, and we were never to leave without preparing to be gone for days. Such was the nature of the sudden coastal storms. For many years to come, there was to be no security other than what we could hope for, create, and call our own.

The old-timers we had met in Homer, curious as to what brought newcomers in early winter, expressed surprise that we were moving to China Pot Bay. The resolve in our voices did little to moder- ate their concern. We took their apprehensions lightheartedly, fearing that it might otherwise reward their doubts in our chances for success.

Old John Waterman encouraged us with comments about the richness of the soil around the archaeological sites where we planned to settle. "You'll see that they're full of shells and bones and so you can see what they ate. That soil will make you a fine vegetable garden; all those shells neutralize the acidity of the soil. You'll see layers of urchin shells; they must have eaten an awful lot of them. Folks here abouts don't eat 'em much, but they're not bad and they're healthy too. You'll have lots to fall back on out there if ya run out of grub." He described how our cove was sheltered from the north wind, and his eyes lit up when he described the bounty of the clam beds there.

We had been especially fascinated by John's stories of the young woman archaeologist from back East who years before had excavated many ancient sites around the bay. "She dug there in China Pot, ya know," he said. "Freddy de Laguna claimed it was a really old village. Some places she dug down nearly twenty feet and was still finding bones and shells and ashes of the earliest people. I reckon that since the Natives had this whole country to choose from, you couldn't go wrong settlin' in the place they chose. Long as ya don't mind not bein' able to get in and out at low tide, you'll find it a good place to live." We liked the things John said, and in the years ahead we became good friends.

Hugh Watson, who had guided and fished, flown and trapped all over the country told us about the name. "Everybody calls it China Pot; it's really China Poot," he said. "It's named after a half-Native fellow whose mother was Eskimo — folks say her name was 'Poot.' Back 'round the turn of the century when the big herring fishery was goin' on in Halibut Cove, there was lots of Chinese workers brought in to work the salteries. We didn't use to call 'em canneries like now, 'cause the fish was salted and put in wooden barrels. So this Native woman takes to hangin' roun' the saltery and the Chinese worker fellas. She has a baby boy which looks Chinese, so they call him China Poot, real name was Henry. Well China Poot takes to huntin' and fishin' over there," he points across the bay, "and 'fore ya know it folks is callin' it his bay, which ain't so unusual I suppose since Ned Jackaloff down the bay has his bay and Dan Peterson has his next to yours there." Wiry as a rail, Hugh looked like he was cut from cloth matched to the woods and mountains. "If you'da come a few years earlier, I guess we would have named the bay for you instead of Poot," he said with a friendly chuckle.

Living deliberately in nature was a theme that came to mind time and again. It is said that if you look carefully into a person's eyes, you can see the soul. I reasoned that if you look carefully into the eyes of wilderness, you can see God and thus better understand your relationship with the great teacher.

We didn't know how to build a log cabin or set a skate for halibut, but knew we could learn. Having been trained as a military officer and having served in a top-secret environment during the Vietnam years, I had been bruised and hurt deep inside by the deaths of comrades in arms and needed the blisters and calluses of hard manual labor to set things right again.

I believed that Diane could do just about anything. We knew clearly what we could do and what we wanted to do, and we were keen to get on with it. We had both worked in wilderness hunting and fishing lodges, I had flown the bush with old-time guides and pilots, and we thought we could create a lodge that emphasized a focus on experiencing nature without taking from it. As explorers we wanted to plummet and map our own inner wilderness while living in a bigger one. We envisioned a sanctuary, a refuge where we could be ourselves, where the land and sea would be our teachers, and where we in turn could share our insights with our clients.

I imagined what we had to do so clearly that the doing of it was mere copying. Nothing is more powerful than a dream held firmly and surrounded in truth. It was on the wings of the dream that ideas sailed into my brain complete with blueprints. No matter that the only substantive plans to help bridge the gap between idea and reality were sketches on the backs of paper plates or napkins. T square, slide rule, winged divider, and compass — the everyday tools of the builder never appeared on the various building sites. In the years ahead we would establish still more inaccessible outposts. The first was to be a camp for photographers and wildlife enthusiasts one hundred miles west across Cook Inlet in what would later be surrounded by the McNeil River Brown Bear Refuge. Another was born on a subalpine mountain lake in the glacier country far above the China Poot Bay estuaries. With a minimum of supplies and lacking electricity and other helpmates, we would develop slowly. The wind, tides, and sun provided the nurturing conditions for our imaginations. The dark and storms, snow and ice tempered our metal. Like a newborn child, the task at hand tugged at the warm breast of the land and sea itself and we believed would thus grow strong and healthy.

The vision of what was possible was founded on unsubstantial ground, in addition to which Diane's parents were opposed to my plans to build a lodge, and their stout opposition was a near-crippling blow. Still, we believed in and strived to articulate a simple statement: we would live completely surrounded by nature, loving it in all its many forms, gentle and savage, and immersed in it, we would be guided by it and create the means to sustain ourselves and the children that we dreamed we would bear in this wild place.

"Gull Island," Alex pointed through the gale to a dark and ominous rock thrusting up from the sea off the port bow. "Good place to get eggs," was all he said.

Living on the Agulowak River the previous year with Native people, Diane had helped collect eggs and tried to learn how to choose those that wouldn't plop a feathered bug-eyed hatchling into the frying pan at breakfast. Some called them strong or "fishy," but we enjoyed them and felt more attached to the land using what it was willing to give. I recalled Diane's youthful delight in describing her adventures searching for eggs out on the tundra with the Eskimo women. I had been concerned that there were many brown bears in the area and some might be searching for the same eggs, equally content with the feathered or nonfeathered delicacies.

When we had inquired for a post office box before leaving Homer, the postmistress had cheerfully chatted with us over the worn counter. The quiet little room seemed frozen in time, a page out of the past. While we talked, Virgo Anderson slid a folded newspaper across the counter and we caught the delicious whiff of smoked fish. "Thought you might like some silver salmon, Arlene," he said. Wearing a friendly smile, Ruth Newman walked in and joined the conversation, greeting Diane and me like old friends. Sharp as a tack, she seemed to know a great deal about the history of the area.

The postmistress unfolded the fine-looking fillet, and we all had a taste. Diane and I looked at each other and smiled. For the next forty years we tried to match that heavenly flavor with our own product. There was and is lively competition between friends and neighbors to see who can produce the best gravlax, squaw candy, canned salmon, and other varieties of the local fish. Since we were newcomers, Arlene, Fred, and Ruth took time to visit, to make us welcome. We told them we had spent all of our savings to buy the unfinished log cabin and old homesites in China Poot Bay. We told them that we were waiting for Alex Flyum to return to help us take our gear over there.

"Word has it that when the weather lays down, he'll be back from Carl Williams's place in Iliamna Bay." Everyone seemed to know Alex and Carl.

Ruth said she didn't have an extra bedroom, but we would be welcome to sleep on the floor at her place at Miller's Landing until Alex came in. We happily accepted. We had had about enough of sleeping in the back of our pickup truck. "We heard that someone bought those homesites," she said with a smile. "I admire your courage wanting to live out there."

Courage, we thought — courage, what courage? We didn't know courage was a part of the arrangement. Courage implies danger; what did she know or imagine that we didn't? No one ever asked what we were going to do over there or how we planned to make a go of it. It was assumed that we could and would do what we needed to do. People were given the benefit of the doubt. Their friendliness was like an anchor to windward. The fact was that people had for many years been withdrawing from the old trapline cabins and homestead sites in favor of living in an established community. We were going the opposite direction from what they were used to.

The postmistress paused to file some letters in the pigeonholes behind the little old-fashioned brass-and-frosted-glass doors that were opened with a little combination knob. We told them about our experiences in the bush, miles from McGrath and Dillingham where I had flown as a bush pilot at Farewell Lake and north of Aleknagik in western Alaska. Diane had cooked in tented hunting camps and in lodges, learning to master the vagaries of baking on woodstoves. We all laughed when Diane told the story of being left alone in a remote camp.

"I picked lots of blueberries around the old Rhone River Roadhouse, which still had dog harnesses, scores of traps, clothes, and equipment left behind by the gold prospectors. To keep away the loneliness I stayed busy making two big pies, then I took a walk along the river, and there were grizzly signs everywhere; they were digging up milk-vetch roots. I sure wished the guys had hunted around camp instead of in the mountains across the river, but they planned to be gone for the next few days. Well, I hurried back, afraid that the baking berry pies would draw those grizzlies like bees to honey. I wrapped the pies in a sleeping bag thinking I'd hide the scent and went to find the gun. Those guys had gone and left me cartridges for the gun all right, but when I went to load it, I noticed this big bend in the barrel; boy was I mad, though I chuckled at what my parents would have thought."

A year or so later, Ma Pearson, a real Dillingham pioneer, had taught Diane to split wood and smoke fish the old way there on the banks of the Agulowak. Our stories were offered to cool the concerns of our friends about our abilities. But the truth was, we knew next to nothing about the open ocean and the huge tidal surges of the bay. We knew just enough and had the confidence to be dangerous to ourselves.

On every frontier throughout history, when newcomers arrive to push the limits of a settlement, the residents take special interest. In a closely knit community, if any one person is in trouble it concerns everyone, and often the collective group pulls together to help the unfortunates who, as often as not, are foolish and unprepared. Staying out of trouble was an obligation to yourself and a kindness to your neighbors.

Our stories in the post office lobby implied a confidence we didn't feel. There was still one penetrating question no one asked and for which we had no answer. "How do you plan to make a living?"

We hadn't a clue. That and a myriad of other uncertainties gave us the hollow and uneasy feeling we felt now as we stared out of the wheelhouse windows at the ugly gray combers.

The cramped wheelhouse was warm and snug against the biting cold outside. Driving south from Anchorage over the mountain passes on the way to Homer, we had encountered really severe cold, heavy snow, and whiteout conditions. Time and again we had nearly driven off the road in near-zero visibility. State troopers warned in radio announcements: "Motorists are advised to let someone know when you are due to arrive at your destination and always carry survival gear."

More than one person had frozen to death when their vehicles plunged off the road into a snowdrift. Although it was warmer here by the ocean than on those high passes, the wind carried a ferocity and penetrating cold that surprised us. We were not emotionally prepared for our surroundings.

Towering mountains thrust up in a wild jumble at the shoreline ahead. Thousands of feet above, they were lost in shredded, foreboding clouds. Elsewhere in Alaska I had climbed similar peaks, pausing in the deep forests, sleeping in the flowered meadows, but these jagged summits looked frightening, hostile, and ever steeper as we approached.

The wheelhouse door at the stern was ajar, and diesel fumes slipped in like an unwelcome visitor. Icy spray whipped over the port gunwale and rattled against the wheelhouse like thrown gravel. The wind had switched the night before from southwest to northeast, bringing colder weather. The big surf we had seen in Homer at Bishop's Beach had been coming from the direction of Augustine Volcano and the open sea. Now it was roaring down the bay from the north, the exhaled breath of many glaciers.

The cramped space we shared with Alex behind the wheel was jammed topsy-turvy with two bunks, a fat little dog asleep on the skipper's pillow, a stove, and a crude galley. Every available space was crammed full of line and wrenches, rain gear and groceries. The diversity of smells was as jumbled as the assortment of seaman's gear.

Annie was tending a skillet of bacon. The coffeepot perked busily next to the chimney where she had wedged it between adjustable steel bars that kept pots and pans from skidding across the stove as the boat rolled and pitched. Galley cooks on every ocean ignore the weather to feed the crew — sailors have to eat no matter what. Diane's eye caught my own as we watched curiously; Annie cut two more thick slices from a sow-belly slab that still wore a nipple and several coarse hairs. I leaned her way as if in affection and whispered in her ear, "cheechakos." We laughed silently at one another and our newcomer status.

I studied the adjustable bars parallel to and a few inches above the stove that held the pot and black skillet in place. They slid along a bar that circled the stove and could be easily released by a wing head screw. I wondered what sort of imaginary bars and screws were going to come into play to keep Diane and me secured to the precarious perch we had chosen for ourselves up ahead somewhere on the far shore.

Unappetizing fumes rose up from the bilgewater that was wildly sloshing around below the floorboards. I hadn't been able to learn much about the *Nanuk* other than that it had worked for many years around Cook Inlet hauling trappers and fishermen, homesteaders and cattle, lumber, trucks, and even other boats. Each had left some essence of its passing, and it was all richly mingled there, the good and the bad in the wheelhouse with the hearty smells of frying bacon and perking coffee.

There was a certain calmness to that moment that has come again and again in a lifetime on Alaska's wild coast. Standing there next to Alex, I felt secure — even proud somehow. It was a feeling of belonging, of having a purpose, of being with others who where simply going about their business with deliberateness and dedication, confidence and determination. Like pioneers all across America's history, Alex possessed a fair measure of self-worth and had created a marketable skill. His personality type was as solid as the frontier itself. He was tough, and although we hadn't seen a measurably gentle side, it was clear that it was there. He was a survivor. His quiet dignity reflected a simple interior beauty. Being in that place with Alex and Annie gave such a deep and mysterious pleasure that still I like to think of it. It encourages deep respect for hardworking people and launches me to dreaming of new friends, new adventures, and new challenges in still more distant places.

Alex was staring intently at the water, studying it — perhaps trying to remember locations of rocks and bars from previous experience. His eyes had become narrow slits after years of staring out of wheelhouse windows. I could imagine his Eskimo ancestors squinting across the frozen tundra looking for a fox or across the ice pack watching for an *oogruk*.

Now on the other side of the bay the north wind was fighting with the outgoing tide, and those two huge forces were at war at the mouth of China Poot Bay. We were going to boldly ride this tired old warhorse right into the middle of the terrific battle.

I had been told that Alex probably knew more about this part of Alaska's coast than anyone alive. He had been born some several hundred miles to the west in the little village of Kokhanok on the south shore of Iliamna Lake, which drains into the Bering Sea. I had heard that the lake is so big it hosts some of the world's only freshwater seals, and like Loch Ness in Scotland, there are persistent rumors of a monster. Alex was almost seventy, and like many people from remote areas born around the turn of the century, his parents had been born underground in a semisubterranean dwelling. These shelters had been made of bones, driftwood, and skins and had been called *barabaras* by the early Russians. His Yup'ik bloodlines linked him to some of the toughest and most enterprising people the world has ever seen. He was a first-line descendant of those who had migrated across the Bering Land Bridge. His ancestors more than ten thousand years ago had moved as one of many waves of migrants from Chukotka in Siberia.

"There's almost always a rip here at the entrance to China Pot," Alex said, pointing with a gnarled finger to the dangerous-looking stretch of water just ahead. Angry-looking whitecaps arched menacingly where the multiple forces clashed. Yesterday's ground swell was trying to cross the bar from the south. The wind's whip and glacial blow was tearing the tops off of the crests. The low sun in the west punched shafts of light through the thrown spume.

Duck hunting as a teenager on the Chesapeake Bay across the continent in a completely different kind of ocean had given me just enough experience to know that this was no ordinary riptide. China Poot empties into Kachemak Bay just south of Gull Island between the bird rocks and cave-studded cliffs on the northeast shore. There the rush of outgoing water as the tide ebbs conjures up images of maelstroms and boats being sucked out of sight. It seemed no place for human nor beast, but there were gulls, guillemots, murres, and harlequin ducks feeding on morsels that were being washed out of the estuary and rolled to the surface.

"It's mighty shallow here at the entrance." The narrow entrance was like a little basket mouth. "Nearly goes dry right here; see, the Fathometer says only fifteen feet of water under my keel. The water speeds up as it comes between the cliff and those sandbars," he said, pointing to the log-littered bars a few hundred yards away. "Don't forget what you're seein' here, 'cause it gets a whole lot worse than this. Ya just don't ever want to try to come or go through here when there's been a big southwest blowin'. When she starts to ebb after a big blow like that," he paused as if to consider our situation, "well, you'll see, you'll see." He pointed to the high cliffs on our left. "You can walk out to the end of the point there from your place."

Pointing again, he said, "I've seen them big ocean-bred graybeards slam into those cliffs and break right over the top of Gull Island back there."

It was an ominous warning. In the decades ahead, we were to come face to face many times with the unleashed terror of crossing that bar. I very nearly rolled our new salmon drift boat in that exact spot. My good friend Tom, a young neurosurgeon, and his teenaged son drowned there and were never found. My own brother pitchpoled his Opheim snag skiff there on the way to town, then miraculously clung to the bottom for an unimaginable two hours in the icy water before his lucky rescue.

The blunt bow of the barge slammed into the rip, and spray flew everywhere. For a while it seemed like the shoe box–shaped hull just couldn't overpower the outgoing push of the tide. Alex's brow was knotted as he studied the water, looking for an eddy or backwater where he might find the going a little easier. It felt like driving a heavily laden truck through deep sand.

Leaving a trail of dark diesel smoke, we surged through the rip and slowly chugged past the tall cliffs at the entrance of the bay we were about to call home. The cliffs were studded with dark caves and hung with strange black cormorants that were like giant bats. Against this somber background, harlequin and goldeneye ducks played impishly among the tumbled boulders at the base of the concave cliffs. Between the cliffs were small beaches, white with foam and gleaming black. Long strands of macrocystus kelp, whose holdfasts had broken loose from the sea floor, rolled forlornly up and down the beaches in the surf.

Once into the cradled arm of China Pot Bay, it was actually calm in its lee. Out of the full force of the north wind and facing south, it was warmer. Ragged clouds raced by and began to thin; the sun tried to probe through a cloud bank in the west. The water changed color from blue to dark-tea green to silty gray. The freshwater coming from the glacial valleys carried a load of gritty glacial flour.

"Swank's Fox Farm." Alex pointed to an old and forlorn-looking one-room cabin above the curving sandy beach. It looked so tiny in its pretty little cove framed with giant Sitka spruces. In the years ahead, we would learn a great deal more about the old bachelor fox farmer

The lodge as first seen by Michael and Diane, deserted and lonesome. Photo © and courtesy of Michael McBride.

who had left many years before and of the ancient archaeological site on which he had built. There was an air of mystery about it.

"Looks neat," Diane said. "It might be haunted." She wore a girlish grin. I could imagine the fun we'd have exploring the place.

"John Swank made good strawberry jam," Alex said. Later we learned that he was well known up and down the bay for his excellent homemade jam. We even found a jar on the shelf of his fur shed before the building collapsed under the weight of the winter snows. The jam was still redolent with the heady fragrance of the old-timer's handiwork and reminded Diane of the smells in her grandmother's kitchen at canning time. Over the years, that small page of fascinating history melted back into the forest floor, and today not a trace is left but for the tall cobalt-colored delphiniums that John planted and that still mark the spot. We learned that, like other early residents on our side of the bay, John had rowed and sailed to town to sell raised fox or trapped furs for cash and to trade things like homemade jams for seed potatoes or perhaps, if he was lucky, for a homemade shirt.

Farther ahead on the port side was a second abandoned cabin whose door hung open like a missing tooth. Even from the channel below, we could see that part of the roof had fallen in. The hole in the roof left a void. Abhorring a void, nature was filling it with snow and rain and the falling detritus of the giant trees around it.

Where were the people who had lived here, I wondered? How long had they lived there? Were there families? What of the children? I was full of questions.

"That little cabin over there above the beach," Alex pointed to the object of my curiosity, "was Brewster's place. He got washed overboard off the fishing vessel *Wilson*. His dad was runnin' it. They were crabbin' off Augustine Island. That was years ago."

A little flock of rock ducks played among the burnished fucus seaweed and boulders at the foot of a little point just off the bow. We motored ahead at a crawl, the intensity of the current abating now that we were out of the narrow entrance. The bay flared wider. Alex was simply feeling his way into it. "Fathometer doesn't read so good when it's shallow."

I wondered what rocks or bars were lying in wait, hidden just beneath the keel in the swirling blue-green waters. From my perch on the bow I could hear the high-pitched little calls of the harlequins: "Thwee, thwee, thwee." Their cheerful song was a musical complement to their playful black, white, and burnt-orange costumes.

Diane had taught biology lab to university students in Anchorage just after we were married. She commented on the richness of the water flowing past that seemed full of seaweed and organic material — the outpouring of an extraordinarily productive estuary.

"Just look at the contents of that water, Michael. What finer welcome mat could be laid for our arrival?"

"Yep," Alex said, "that water is as rich as any soup; this little bay has more clams and mussels, ducks and gulls, seals and salmon than any place I know. Leo Rollins shot a big brown bear just over there a few years back." He pointed across the estuary to where a band of cottonwoods wound their way through the spruce, apparently following a creek.

"Are there salmon in that creek, Alex?"

"Yep, silvers and pinks. On good years, lots of them," he nodded.

"Well, that's where we'll get the salmon to smoke next fall for Ruth and our new Homer friends," Diane said.

"Nig and Nora Lippincott came up from Seldovia years back to trap that valley, did pretty well too, I think. Nig told me about a pack of wolves there." Alex was getting downright chatty. Hugh Watson, the old guide we'd met in town, said there were moose among the willows in that valley and that we'd probably be able to get our winter meat there. "Clem Tillion takes his commercial fishing boat, the *Ram*, in over there, lets her go dry against the bank, shoots ducks, and jars enough for winter. Lotsa black bear, 'n' river otter too. You picked a good place to live, alright."

Those few words meant more than gold itself to us, and we would make them our own.

"Just ahead now," Alex said as we passed an especially beautiful little cliff so close and sheer that someone standing on top of it could have tossed a spruce cone down into the open barge. It rose straight up for fifty feet and dropped directly into deep, swirling water. There were small wind-stunted spruces growing from the rock face that had been racked by a thousand storms and would have been the pride of any bonsai gardener.

The delicate layering of the rocks there, like curled pages in a colorful book, seemed to spell out in some forgotten hieroglyphics a clear message of welcome. We had the sensation that it was a sign from the ancient residents who were preparing to entrust this place to us, happy perhaps at the thought of new arrivals wanting to make a home. The first of them had probably arrived with similar hopes and fears. As surely as God made barnacles, there had been a first people to arrive here, and they may have been individuals similar to us. What did they look like? Was that first person a solitary hunter foraging for food or a family group pushing the boundaries of society as they knew it to exist thousands of years ago? How did they feel having left friends and family behind? As we hoped for a new life, we felt the shadows of the little children and the old people. We imagined skin boats on the beaches and smoke curling up from their shelters. We could see it all clearly somehow.

"The ancient people who lived here," I said to Diane, quoting from reading I had done about them, "lived in a world of extraordinary interpersonal activity. Each tree, codfish, rocky point, and sparrow not only had spirits to be dealt with but were actually seen as people with whom one communicated." Because they had lived there for thousands of years, their presence was tangible.

Later we would find the muted evidence of their passing and even their bones peacefully at rest, slowly returning to the soil. We were contentedly one with those people from the very start, but only because they welcomed us.

On our first visit to this little cliff the previous spring, Diane and I had walked out to the point from the old cabin and fallen in love with the wind-shaped trees and gnarled roots curling over bedrock. While the dark forest places were still wrapped in early spring slumber, the south-facing wildflower cliff gardens were bathing in the long afternoon sun. There were bright yellow cinquefoils with the smooth silver velvet under their leaves and pale milk vetch, which bled milky sap when plucked for a bouquet. There were shy little violets, boreal Jacob's ladder with colorfully accented stamens, and a dozen others.

Perhaps best of all here was a majestic view of the snowcapped mountains and stately glaciers above them. There was even a huge waterfall pouring down from a high cirque to the valley floor beyond the estuary. We could see most of the salt flats from the point. Beyond the barrier sandbars that protect the estuary from the onslaught of the great oceanic waves was a clear view of the active volcano and the open inlet. The bar was in sight too, constantly reminding us of our insignificance.

We had been more than entertained; we were captivated. We were fascinated by the Monet-like softness of dancing light, yet there was boldness of color in the play of sun and cloud on sea and shore.

We had fallen under the spell of the giant trees, the singing of warblers and song sparrows among the alders, the dappled light dancing on the forest floor, the sea cliffs, and the broad estuary beyond. We had explored, filled with amazement that anyone could ever actually "own" such majesty, such imperial beauty.

Every point, cove, and bend in the trail held a new discovery and unfolded a new vista. Here was a sea otter asleep in the kelp where it had wrapped itself in the fronds to keep from being carried away by the current. Cobalt blue jays, saucy in their indiscretion, swooped down on our picnic lunch and snatched tidbits, then chattered amiably among themselves. There was a herd of seals asleep on the sandbar, clearly visible from the cabin window. And eagles — there were eagles everywhere! We had seen a dozen or more perched on driftwood snags, high in the spruces, down on the sandbars, scavenging on exposed mussel bars, and there were nests all along the shoreline. We saw moose hair where it had snagged on a low branch and claw marks on a spruce where a bear, catlike, had raked the trunk as high as I could reach, marking its territory. We had seen puffins and woodpeckers, plovers and grouse, hawks and owls, thrushes and wrens. The abundance and variety of birds alone was staggering and told us that there was a less apparent and even richer biota that supported them. We had only a brief glimpse in the intertidal area, which promised to be the richest area of all. Old-timers in Homer had said, "When the tide is out, the table is set."

There was such promise here, and everything was familiar somehow, as if we had known it of old. It was to be a home for a lifetime. Our axis would pivot on our shared energy and on the generosity of the land.

There was so much to take in at first — we were giddy with the newness of it all — but it was the great trees that perhaps most impressed us. Among the massive-limbed giants we found some that were more than fifteen feet in circumference, and we knew that they had seen the comings and goings of centuries of history. We felt tiny and insignificant in their presence. When our son, Morgan, was born a few years later, we would name him for the sacred groves of trees near the sea so dear to the Celts. He would be Morgan — in Gaelic, "one who lives in the forest by the sea."

Above it all there was a magical quality, like a serenely waving silk scarf, a perfection of Zen harmony, a balanced union of yin and yang, sea and shore, mountain and salt marsh, wave and cloud, towering conifer and luxuriant beach grass, all blended in a terse dynamic tension as if a conductor held a poised baton. The great tidal pulse, rising and falling twenty vertical feet and more in six hours, called to mind my love of Bach and his punct/conrapunct or point/counterpoint. The tides were indeed like a gigantic metronome, and Johann Sebastian was the conductor.

During that visit the previous spring, when we had decided to buy the land, we had marveled at the teeming intertidal life side by side with the sun and salt-bleached bones of driftwood, which lay polished on the sand. A few strides farther into the forest the sphagnum moss, lush ferns, pyrola, hemlock, and false azalea nodded in the damp green quiet. The leaves, trees, grasses, flowers, and blue sky shouldered up to giant trees and adjacent snowcapped peaks. Sorceress-like, it cast an enchanted spell over us, and we thought we'd burst from the simple joy of the present and the promise of all that lay ahead.

We found a cathedral surround in the forest. There were massive-trunked monarchs with great drooping branches laden with mosses. The place was within reach of the sound of the surf. Softness filtered down from the needles high above like light through stained glass. Slanting crepuscular beams penetrated the darker places as if an altar boy had just passed swinging a smoking censer. A prominent bear trail crossed a bald knob of glacier-scoured bedrock that bulged through the sphagnum moss like a tensed bicep. Standing reverently on it I imagined how, for centuries, people, bears, wolves, wolverines, coyotes, and others had walked across it and paused, senses alert, looking, listening. In that spot I could feel a warm current of energy starting at the top of my head and flowing out through the bottoms of my feet and directly to the center of the earth. In such a place the veil between this world and another became thinner.

This place and this feeling began to define us as we were to define ourselves, as teachers define pupils and pupils define teachers. Living apart from a traditional community, we were naked in our innocence and naïveté. Yet we were bolstered by our optimism and confidence that this place above all others was right for us and we were right for it. It was to become a union so deep and intense that it sustained us in times of great adversity.

Even during that brief first visit, it was apparent that there was, among the trees, rocks, mountains, streams, waterfalls, and rivers not just a symphony made up of their individual and distinctive voices but an elemental sound that combined them.

Perhaps our most sublime achievement and the focus of our spiritual direction would be to meld our own heartbeats into synchronicity with these rhythms, like a nursing child's breath and pulse blend with those of the mother, making them one.

On that sun-kissed day of discovery a year before, the soothing dance of the waves on the beach spoke of the fleeting nature of time. The distant drone of the de Havilland radial engine drew nearer and pulled us back to the present. Our pilot, Bill DeCreeft, banked his wings for a turn to his final approach, lowered his flaps, and skipped across the shallow water for a landing.

The steel forefoot of the old *Nanuk* ground into the rocky beach with a ferrous crunch, and the warm recollections of that spring day vanished.

The heavy gangway creaked and groaned, lowered by greasy cables. Alex idled the engine in gear to hold the bow solidly in place, and we quickly unloaded our few supplies. Diane did yeoman's work with the heavy boxes of books, tools, and clothes. Alex and I wrestled the heavier things, and in short order we were finished.

Alex didn't say much; there wasn't much to say. He raised the gangway, eased into reverse, and was headed out with the tide before we had a chance to consider what was happening. In a moment the *Nanuk* was out of sight around the corner of the cliff. The reassuring sound of the diesel faded, and we were left standing there, alone. Completely alone.

The tide was falling; it would quickly rise again and swamp our gear. As if to set the tone for decades of hard work ahead, there was a lot to do and not much time in which to do it. The heavy gear had to be carried over slippery rocks, up the low bluff, and into the old cabin before nightfall. We numbered our possessions. There were crates of food, wooden Blazo boxes of fuel, books, a tiny flashlight battery–powered record player with twelve-inch records, and a lovely old Martin D12 guitar. The cherished Blazo boxes would be valuable for kitchen shelves or for tool storage, and one later became a crib for the firstborn child. Diane proudly carried up her galvanized bathtub, our great concession to luxury.

The log cabin had stood deserted and unused for years. Although it was sound, it had seen the abuse of time and an assortment of passersby. Hunters and fishermen had left a trail of litter. The stone fireplace was heaped with half-burned garbage and charred bones. Empty cans and broken bottles littered the floor. Half a sheet of broken plywood covered part of a window and swung by one nail, thumping now and again as the wind rose. There was no wood for a fire; it was dirty and ghost-like. It would soon be dark.

I found candles and lit one, then another and another. "Just tonight, my dear," I said, "I know it seems wasteful, but let's light them all if only for a few minutes; this is an important celebration, you know."

I dragged in branches from the forest, and soon the crackling fireplace and bright candles made us at home. The old Monarch wood-stove slowly produced a kettle of hot water and there was tea with honey. We sipped it slowly and tried to soak in what we had done.

The fire wouldn't take the chill out of the room. Inside wasn't much warmer than outside, so we each dug out another sweater. We were surrounded by forces so much bigger than our own. We felt very small and very alone.

We couldn't recall which crate held the Coleman lantern; pots and pans were at the bottom of another box — it had been a long day. Pilot Bread, Spam, Darigold butter from the little round can, and more tea would be good enough for dinner.

"Civilization had its beginning around an open fire." Diane liked it when I told stories or recited long poems. "There at the fireside was the family gathering. Children learned, people shared and loved, life attained meaning. If you trace the origin of the Latin word *focus*, you find it related to the word *fireplace*. For here the family group found safety against the darkness and uncertainty of the night, shared warmth and food, broke bread, enjoyed companionship, friendship, laughter, and tears. If you speak of a person that they are homeless, they are absent from the pleasure and comfort of such a place. I am glad to be here with you, my dear, beside this fire, as we begin to create such a place, a home."

I found a thermometer in our gear; it read twenty-two degrees inside. Diane playfully sat on a wooden Blazo box with her feet in the oven of the woodstove. Even that didn't do much good for her frozen tootsies, but it did make us laugh. The forlorn cabin had been abandoned for years and was not only cold soaked but spiritless.

Guided by Robert Frost's poem "One More Brevity," we decided that our "last look should be taken outside a house and book." We went to see if we could find Sirius, the bright Dog Star between the dissipating clouds. We did, and we also caught a glimpse of the dog's hunter, Orion, rising over the mountains. "Look, Diane, there's Aldebaran, the eye of the bull in Taurus, above Orion."

She then replied, "Look, honey, there is Ursa Major with Merak and Dubhe pointing to Polaris." The North Star was only several degrees shy of being straight overhead.

The cold drove us back inside the cabin, where we stood shivering, hand in hand. The stars were looking in through the window; the twinkling light of the candles was reaching out to meet them.

The only heat was next to the stove. "Wanna know where the warmest place in the house is?" she asked. "It's in our sleeping bags." She gave me that cute come-hither smile and scampered up the stone ledges of the fireplace to the crawl-space loft above. I tossed up the goose-down bags. I blew out the candles and stood there for a moment gazing at the estuary and the dim light of the stars across the water.

"Michael?" she called softly. I bounded up the ledges, and soon we were snuggled more closely together than we ever had been before.

In the years to come, climbing those stone ledges marked the end of our days. Some of those ascents would be heavy with exhaustion and leaden with despair, others light with celebration of some small achievement. But on that night of fulfillment there was only the sound of our own contented breathing blending with the sighing of the wind in the spruces and the music of the crackling fire on the hearth. Our search had ended, our journey had just begun. Never were Diane and I to have fewer possessions or greater wealth.

Miaka –
"Sensei San" Michael's
First Naturalist Teacher
Note Tadpole Can /
Insect Box

Michael as a boy while in Japan with Miaka, his first naturalist teacher. Photo © and courtesy of Michael McBride.

CHAPTER 2

How It All Started

*"How happy I am to be able to wander among bushes
and herbs, under trees and over rocks; no man
can love country as I love it. Woods, trees and
rocks send back the echo that man desires."*

—*Ludwig van Beethoven,
on his Symphony No. 6, September 1812*

Among a collection of old family photographs is one my mother took of her firstborn son at about six or seven. I am wearing shorts and kneeling in grass beside my teacher, a young Japanese medical student. I have a small box under my arm, and it contains insects and butterflies impaled on pins in the typical collection style of the time. Under Sensai San's tutelage, I came to know the beauty and complexity of the insect world brought to life under the waxed-paper parasol of Buddhist respect for all life.

These early natural history lessons gave me a deep and abiding appreciation and respect for the natural world in all its forms. My friend-teacher took me to the shore where the clear river flowing past our house met the sea. At dawn, we pushed our little boat from shore with the octopus fishermen who had built their craft without a single nail. The wonders of the undersea world were revealed to me as we peered over the side into darkened glass-bottomed looking boxes. There we watched tako the octopus and puffer fish and a child's delight of marine forms gliding along on their submarine business. As for young children anytime, anywhere, the power of youthful experiences can shape whole lifetimes. As Harvard professor E. O. Wilson, who would become my guide in later life, wrote, "A child comes to the edge of deep water with a mind prepared for wonder."

We were among the first American children to enter old Nippon just after the war. It was a remarkable time in the Land of the Rising Sun, and we were at the very leading edge of a tiny piece of history. Marco Polo in the court of Genghis Khan was no more dazzled and intrigued than was this impressionable young boy in the midst of the sights, sounds, and smells of the exotic Far East. The opportunities and experiences afforded this young fellow were to deeply color his sense of the world and his place in it. I see a look in that young fellow's eye that portends the story of why I decided to spend the rest of my life immersed in the natural world of Alaska's wild coast.

Our ship took the great circle route eleven days enroute from Seattle to Tokyo, and it would have taken us within sight of the Islands of the Four Mountains in the Aleutians. Standing at the port rail and gazing across the fog-shrouded sea from a heaving deck, I might have caught an unknowing glimpse of a smoking volcano. The people who had prospered in those remote islands for ten thousand years had advanced mummification skills to a degree of sophistication approaching that of the Incas and Egyptians. Towering above its neighbors and belching smoke and cinders, Chuginadak Island cast a mysterious net over me and stimulated a lifelong interest in remote, unusual, and little-known places. After a lifetime of adventures around Alaska and around the world, as a gray-haired fellow I would be dropped off alone with a single folding kayak on that very island, at last coming full circle. This is the story of what happened in the interim.

In 1966, having completed four years of military training that ended with an officer's commission and a fast track to Vietnam, I chanced to study in the library at the University of Illinois, and there on the steps I met the love of my life, the charming blue-eyed lass who would give birth to our children, who would take on the mantle of her ancestors who were pioneers and homesteaders and carry that heritage north to join me in Alaska. She would add an unanticipated and delightful dimension to my bachelor's world and would later climb with me up a steep mountainside of obstacles that led us beyond the end of the road into the wilds of coastal Alaska.

Perhaps those bright-blue eyes set between her blonde pigtails contained some sparkle from her Cherokee great-grandmother, but whatever it was, it was good stuff. Very good indeed, and before you knew it we were playing footsie under the coffee table at the Turks Head. Before I knew it, I was taking the first train ride of my life on Illinois's North Shore Railroad to meet her parents.

Diane had been raised in a culture dramatically different from my own. My blue-eyed girl came from a long line of pilgrims going back before the Revolutionary War. Her great-, great-, great-grandfather preserved for us the frontier of Kentucky in his book *Cyrus Edwards: Early Days in Kentucky*. He described the woodland

Indians who ranged freely through the forests while the dwindling smoke from their campfires across the river signaled the end of their thousands of years of supremacy. Many of Diane's family went on to great distinction: ambassador, newspaper editor, missionary, artist, entrepreneur, lawyer, author, teacher. One colorful fellow went west during the California gold rush. As he traveled, he married a proud Cherokee woman who honored my future wife and children with the blood of the first Americans. Diane inherited many qualities that made all the difference in whether we were going to sink or swim in our attempt to build a wilderness lodge in a roadless place.

My Celtic ancestors by contrast had so recently crossed the Atlantic that they still had salt on their eyelashes. The coast of Donegal northwest of Sligo held little opportunity for Hugh McBride's family before the turn of the century. So with a potato in his pocket, his father sold everything he had, which wasn't much, gathered his meager savings, and bought passage for the little family to America. My dad's father told and retold of his adventures as a child playing on the deck of the sailing ship. Apparently Dad felt the tugging of his grandfather's genes and kept moving. The tough and frugal farm life in Wisconsin was not for him, and an army career looked like a way out of a hardscrabble existence.

It wasn't lost on me that my girl came from good stock. I thought we might just be the kind of pair to take on the homesteading challenges in the Alaskan wilderness that I envisioned. It's not socially acceptable to examine the teeth or feet of one's beloved as a horse buyer might, but it occurred to me that her gene pool certainly offered odds that she was tough enough for what I had in mind. It would take years to find out what kind of mettle she was really made from. I was to learn soon enough, though, and then bragged about her. Still, her conservative-minded Midwestern parents hadn't much in common with the adventurous and worldly members of my own family.

My adventure-oriented and free-thinking family swam in a different pond. Trouble was, I was a soldier headed for The Last Frontier that had the dubious distinction of being the jumping-off place for military men going to Vietnam. I had no business getting mixed up with a young woman. She was only twenty — a junior in college — and I was an about-to-leave-town military guy. This was crazy!

We were the last colonists in our family histories that moved west from Europe and west again from their immigrant landings on the East Coast. As our old friend from Motel 6 Tom Bodett says, we decided to go "as far as you can go without a passport." Others have said that Alaska was "America's last gasp of Manifest Destiny." In the years ahead of building our home and lodge, there would be times that we were gasping — that is for certain!

When first I saw Alaska I knew that I would never leave. Assigned to Anchorage, in 1966 I quickly saw that the military people usually kept to themselves on base, which was occasionally locked down in that Cold War era. Determined to meet local people, I got a second job nights and weekends in the freight yards.

One day, my boss asked me to help move a piano across town. "No problem," I assured him. Arriving at the designated address in my big truck and coveralls, I was surprised to find that the apartment was in the loft of a three-story complex known as the Boyco Building. It was just a block from the heroic bronze statue of Captain James Cook, which was perched on the edge of the bluff overlooking the inlet named for him. He had a stunning view of twenty-thousand-foot Denali. White beluga whales were often seen chasing the salmon headed up Ship Creek, which flowed between the military base and downtown. Fran Lotsfelt, the single woman who lived with that third-floor view, wanted her piano moved to her new address. The stairs had been rebuilt since the piano was taken upstairs and were now so narrow that I decided to rent a wheeled crane and take it out through the window. When I commented about the scoring of the Chopin études I saw on the piano's music stand, she invited me to join a gathering of friends the very next night. As a newcomer to Alaska, and a military man at that, I felt I was the luckiest guy around to have found a conduit into a group of local people. Fran was one of the few who had been born and raised in Anchorage and came from an old family there. The fates must have had their wits about them, for the next evening at her party I met a former Alaska Railroad employee and man-about-town, Frank Buske.

"Say, Michael, I'm driving down to Kachemak Bay next weekend, would you like to go with me?" This was music to my ears. Here I was, a Johnny-come-lately to Alaska, and as a military guy among locals, something of an anomaly. Here was the opportunity to really get to know a local person by taking a trip to another part of the state. When we drove over the hill with its commanding view of Kachemak Bay, I was as certain as any impetuous young man is ever certain about anything that this was it.

I knew at a glance that I had found the place to build a life. The childhood Zen garden in Ashiya-Gawa was a mirrored reflection of this grand seascape before me. The moss-capped stone walls surrounding the family enclave and bamboo grove at Miyanoshita became the great glacier-capped peaks and their shoulders of coniferous green. The foamy shoreline and many river valleys winding up into the foothills offered the same enchantment that I felt in the ancient garden beside the river where it flowed to the sea. The hanging alpine valleys and hidden coves and bays promised a lifetime of exploring and adventures. Here I could learn another language and become conversant in the nuances of nature. Ashiya-Gawa and Homer couldn't have been greater opposites in culture, society, and nature itself, but somehow the two were fused and blended before me in a kind of transcendental magic. I was immediately and irreversibly captivated by its spell. Out of the mists over the glaciers I was offered the contentment of knowing clearly that when the time came to breathe my last, it would be in this blessed place of departure.

To this day I clearly remember extending my arm and doodling a path down the far site of the bay and saying, "I'm going to spend the rest of my life right down there somewhere." The doodling finger didn't, however, tell me how I would make a living. Since truth is often stranger than fiction, it happens that I was pointing to China Poot Bay.

We met Clem Tillion on the end of the Homer Spit at the boat harbor and climbed aboard his vessel, the less-than-glamorous wooden-hulled boat named the *Ram*, and headed for Halibut Cove by way of Gull Island. The tidal passage between the teeming bird colony and nearby cliffs only galvanized my certainty that this was the place. I felt like the little boy who found the oil lamp in the gutter and, wiping the mud from it, caused the genie to appear. It was as if I were being given the wish I had dreamed of for most of my life. At that moment, the property that would call this bachelor's name was only a mile from the keel of Clem's boat. I hadn't yet walked through its forests, but, inaudibly, it whispered to me.

Halibut Cove was and is a picture-postcard idyll of what someone like me hoped to find but hadn't dared dream of. Without roads or means of access except by boat or floatplane, it lay on the edge of a vast glaciated wilderness whose spruce-clad verticality towered above everything and everybody. This little nest of a community was set between sea and mountain and blue rivers of glacial ice like an emerald jewel in a Tiffany setting. The place and its few inhabitants graciously took me in, and I was completely, blissfully at home.

Now perhaps was the time to consider adding a woman to my equation for the future. I asked her, she said yes, and I arranged our marriage ceremony when she came to Alaska during her Christmas vacation from college in 1968. She didn't know a single person at the wedding, which I had arranged in its entirety, but she was to become lifelong friends with several of that group of strangers. Anchorage was blanketed in one of its heavy ice fogs when every branch of every tree and every object large and small was encircled with a fairyland of delicate frost. It subdued all sound, and in that quiet hush a certain magic took hold of us and whispered, "If you think this is grand, wait until I show you my summer." We had come for the North and were soon to discover the West.

Having seen Halibut Cove with Clem, it became clear that people could live happy, productive, and successful lives in a roadless place like that. Growing up as the child of a career military man I had never imprinted on a place, although my years in postwar Japan gave me the attitude that living outside of the contiguous United States would be easy. This place was certainly unlike anything I had seen in America. Something told me from deep in my belly that at last I was home! Halibut Cove told me clearly that I was not running from, but rather running toward, a new and reinvented life. Although battered and bruised deep inside by what Vietnam was doing to my classmates, my friends, and my country, here was the place of healing.

When we had lived in Anchorage in 1968 before coming to Kachemak Bay, we lived next door to John Delaney and his one-room log cabin. We had rented an old house built for workers during the early days of the Alaskan Railroad. Sharing the basement with two ladies of the night, we were occasionally awakened by men banging on our front door instead of their door to the windowless basement. Such were the crazy days in Alaska's biggest and wildest "village," as the Native people called it.

At the Fourth Avenue dive the Montana Club, I introduced Diane to Last Frontier, rough-around-the-edges nightlife. When the very funny Mr. Whitekeys, Doug Hagar, had finished his signature tune of the day, "Elephant Doodie," with his band, the Spam Tones, I reminded him that it was Diane's birthday. He engaged the rowdy crew in singing "Happy Birthday" to her in as loud and boisterous a rendition as anyone ever heard. When the crowd set to dancing it was interrupted by a woman screaming at the top of her lungs from the back of the room. She had fallen through a rotten place in the floor and was dangling there, suspended by her arms and upper body. Diane and I still laugh at the image of the guys in the basement shooting craps on the floor, looking up to see a woman's pink undies and legs dangling through the floor and wildly flailing the air.

When I made the decision to live in Alaska for the rest of my life, I set my sights on acquiring the skill sets I thought I would need. But who were these people scattered all across those millions of acres of Alaska's wild country, what skills did they possess, what character traits enabled them to be survivors in that vast outback? Did we have the mettle to make it? At the time, most of the people who lived in the bush, away from towns and villages, hunted and trapped and perhaps fished commercially in the summer to make money to bring back to the homestead or remote cabin site. We had met some of these people, and they seemed to possess a dazzling array of talents that we did not. Could I — did I — want to learn to be a heavy equipment operator, skilled with cutting torch and welding rod? It seemed that many or even most of these remote types had those skills and a wheelbarrow full of others.

I began to acquire skills that I thought would serve us best. By never taking a day off until I had accumulated weeks of leave time, I began guiding for a big-game hunting outfit and simultaneously earned a commercial pilot's license flying in bush planes and was taught by hunting guides/pilots who knew the country. At one point in that training, I was asked to fly a Cessna 185 on skis to Shishmaref at the Arctic Circle to resupply polar bear hunters who were hunting on the ice in the Bering Strait.

Our marriage was still new when we crossed a stormy bay on the sharp gray edge of winter to begin a new life. Towering snowcapped mountains and glaciers hovered above us as we stepped off the bow of the Native elder's boat and steadied our feet on the broken rocks with as much reservation as excitement. We may as well have been standing on Mars. There were no neighbors for miles in any direction; there was only a deserted log cabin whose construction had not been completed and that had never been lived in. There was no electricity and no running water, or radio, and there would be none for years to come. There was no access to a hospital, no means of communication with other people, and there was no means of making a living in such a place. As such, it wasn't much different from what first arrivals experienced on any frontier throughout history. There were some pretty scary question marks here. At that point, we were long on formal education and short on real-world experience.

When we moved into the unfinished and deserted log cabin in November of 1969, we didn't even have a boat. We had arranged a pickup, but without a radio, if either of us had gotten hurt, we had no way to get to town. Eventually we acquired an eighteen-foot plywood dory with a hole cut in the floor for a sea safe well where a twenty-horsepower Johnson outboard motor was mounted. Those unaccustomed to dories were scared silly at the way it quickly jumped off center and pitched wildly to port or starboard at the least provocation. Diane was no fan of the boat she called "Too-Tippy." More than one person was nearly pitched over the side with its surprising capacity to lunge suddenly to port or starboard.

Exposed to one hundred miles of open ocean, we continued to cross the exposed bay regularly in all kinds of weather, facing all manner of difficulties. There were conditions when the space between near shore and far shore might as well have been the distance between galaxies. It happens that those who leave one shore are not always the same as those who arrive on the other side. Baggage of many sorts — real and imaginary — is lost and found when crossing open bodies of water. A whale may jump, a wave may slam the boat with a heavy crash, salt water may sting your eyes, or an icy trickle may find its way down your throat, across your belly, and then puddle in your navel. You may experience dread or exuberance, laughter or tears. You may end up in an entirely different place than you expected and not even know how you got there, and of course you might get lost in fog. We laughed at the quip, "Don't let being lost spoil the fun of not knowing where you are." More than once we wished we could laugh when we felt lost among our difficulties.

In the shelter of our little cove facing south to the broad salt-water estuary, we reached for our fullest potential. Our first challenge had been simply to get to this place that we wanted to call home, and the next was to survive for a while. So we began to build a shelter — a nest and a home — with the hopeful vision that over time what we built would actually be able to accommodate visitors who would give us money for the privilege of sharing our home. We were empowered by words we had heard somewhere: "Find a place that is lovely, build a fine home, invite people in; love will occur." The tall snowcapped peaks surrounding us looked tolerantly down upon the curling smoke from our woodstove. That ascending smoke drew up with it the hopes, dreams, and aspirations — not just our own but those of others who recognized the dynamic tension and natural balance.

The yin and the yang — the dramatic opposites of the tides and the seasons of brilliant summer days and long winter nights — struck a chord that resonated in the hearts of those soft enough to hear it, strong enough to feel it and become one with it. Just as some musical forms resonate with some people and not with others, what we heard and felt did not work for everyone.

Schoolbooks had taught us about Maslow and his hierarchy of needs — shelter, food, and safety — and we were on the first rungs of that ladder. The nearest source of water was a spring that was a long walk through a dark forest of huge trees on a narrow bear trail. In that towering spruce forest, there were healthy populations of bears and wolves who left their scatty calling cards regularly along that trail, reminding us of their claim to the land. Winter was staring us in the face. We were without a chain saw, and although pioneers had survived and prospered in this country for a long time, they at least had crosscut saws and knew how to sharpen them. The Native people whose land surrounded us had

44

learned to heat their semisubterranean homes with great ingeniousness and possessed none of what we would call the "necessities." Without a neatly stacked pile of dry firewood, we broke branches from the great spruce trees and gathered fallen wood, but neither contained the resins that could make the top of a woodstove hot enough to perk a pot of coffee. Perhaps worst of all was the fact that we had no security, nothing to fall back on, no one to teach us, no jobs, no money, and no prospect of finding them. There were threats and dangers, both real and imagined, not the least of which was facing the open ocean with an inadequate boat and minimal knowledge of the rocks, reefs, sandbars, and riptides — let alone what to do if the engine failed us. This uncertainty, however, created a curious state of mind that people around the world have known since time immemorial in the face of similar obstacles.

We had a dream, a bright and enduring dream, and there was a fresh north wind blowing to fill our sails with resolve. We believed that we could build a home in this wild place, raise children here and homeschool them, tend a garden, feed chickens, fish a salmon net, and hunt moose in the hills and mountain goats in the alpine glacial valleys. There were several kinds of clams in the beds just beyond the front door, and streams that saw thousands of red, silver, and pink salmon all within view of our log cabin's kitchen window. We had been told that prized king salmon could be caught year-round below the cliffs that guarded the entrance to the bar, and the soil looked like it would grow potatoes and root crops. I had a wife that I liked and admired, and she, poor girl, had a husband that would have to do until a better one came around. Diane called my attention to the ads in the *Alaska Sportsman Magazine* touting that "mail-order brides make wonderful wives." "You'd better be good to me," she would say, shaking her finger, "or I'm gettin' me one of them-there mail-order dudes."

Our simple goal was to live a quiet life completely embraced in the arms of wild nature. Throughout it all, our enduring dream and youthful energy — like a philosopher stone turning lead to gold — calmed the uncertainty, made more manageable the dangers, transformed worry to optimism.

It was clear after that first winter of getting to know the place and each other that we couldn't stay there; there simply was *no* money to buy more food. Diane had eaten up the two cases of military surplus C-Rations while pregnant. We were obligated to the $286 per month in mortgage payments, but we had no income. We had no boat or motor or means for getting either. We would simply have to move to town and look for jobs. Looking for jobs in Homer in the winter in those days was like looking for a snowball in hell. We envied those who had regular jobs and a regular paycheck. It looked like the best we could hope for were menial jobs, but we had little chance of finding one, let alone two.

I made Diane laugh as I whined, "You mean I have to go to town and get a real job?"

While teaching Diane how to make her perfect moose meat gravy back in Homer, local legend and pioneer woman Grandma Wally told her about some abandoned cabins up off of Pioneer Avenue. Across the street and up the hill from Erling Broderson's liquor store there were five sorry-looking cabins in what had once been a little cabin court. They were in sorry shape, all right. The best of the twelve-by-sixteen-foot cabins had a broken door dangling from rusted hinges, broken windows, and a nasty mattress on the floor that was apparently used by stray dogs in the neighborhood. Someone in the sleepy little town was keeping pigeons, and they came and went through the broken window and perched on top of what remained of a shelf. "Eureka," I thought. I would offer the owner a deal wherein I would trade free rent for my carpentry skills, such as they were, to fix up the derelict cabins and get them rented.

It was not a pretty sight, and I was hard pressed to think of bringing my darling girl into this mess. There was no stove, but George Bishop told me where I could salvage one. I found a truck-tire rim at the dump for the stove to sit on. A screw valve dripped fuel oil into a burning chamber about the size of a five-gallon bucket, and that drip-drip was adjusted up or down to drip slow or fast. It was dangerous as hell, but it was what we had and we made do.

Someone suggested that I track down Doug Collins. A small town at the end of a long dead-end road meant that it wasn't an insurmountable challenge. In his role as the uniformed chief of police in the little one-horse town, he was quite the figure. He was the chief of police, all right — he was the only police officer. We called him "Deputy Dog," as he looked a bit like that cartoon character, with a belly that flopped over his belt buckle and an accent right out of Alabama. He was a good man and in possession of a lighthearted sense of humor. His ample wife, Bobbie, took care of a passel of kids that tumbled in and out of the windows and doors like acrobats. In his free time he was the pastor at a local church. It would be fun to look back at the police record during his time as chief — I could imagine some very funny situations.

Finally, Diane got hired on as a waitress in a run-down, smelly bar, and I got a job on the "slime line" at the cannery. I thought of the futures our parents had in mind for us when they packed us up for college, of the checks they wrote for that schooling, and of the bright hopes they held for their children's futures.

One by one we fixed up the little cabins, gave a facelift to the shared bathroom and shower, and helped get five of them rented. Living rent free in one of the old cabins and sustained by moose and other meats from my fall hunts, we thought we were doing pretty well. Friends shared generously from their gardens and fishnets, and more importantly, they shared a wealth of information that proved invaluable. We carefully saved every tidbit that came our way about how to live off the land instead of being driven off it, the fate of many who did not learn the dance.

Diane was happy to leave the smoke-filled bar and the angry ways of her ill-tempered boss when she got a better job, this one at the cannery. She heard that they were looking for a lab technician and she applied. She didn't really have the qualifications but did a fine job. Several times each day, she went down to the processing floor to get a sample of crab or shrimp or fish from people working on the respective lines. Of course, she needed only a tiny piece for a petri dish culture, but they insisted on giving her a generous bagful. They told her not to return it to the line after she took her sample, so the fellow she slept with was now well fed on fresh seafood.

My young, upwardly mobile wife next got a good job as the assistant to the local dentist, Dr. Marley. He was a pilot too, and they regularly flew off in his Cessna 170B to remote gravel strips to provide dentistry in Native villages. Cleaning the snow from the airplane, Diane fell one day and suffered miserably for years with back pain without telling her boss, who was our good friend and ally and helped us regularly in many ways. I was spending most of my time in China Poot Bay among the two-by-fours and tar paper purchased with Diane's paycheck. By now I had built an addition to the log cabin, and the first clients in 1970 were Fairbanks open-cockpit Alaskan aviation pioneers Sig Wein and his chief pilot, Frank Whaley. Wein Consolidated Airlines would eventually become Alaska Airlines.

Hungry for any kind of work, I put an ad in Lucille Billings's little ditto machine–printed newspaper. The clunky hand-cranked machine had a shiny cylinder to which was attached a waxed paper on which the news had been typed, "hunt-and-peck" style, on a manual typewriter. We chatted as I cranked the handle. With each rotation, a copy fell into a pile. Before we knew it, we had printed off a few hundred copies of what passed for a newspaper. My little ad said, "Carpenter with own tools and truck, will do anything, anytime, anywhere, call me." Trouble was, we had no phone and couldn't afford one, so Dr. Marley let Diane take calls for my jobs in the dental office. Imagine such kindness on the part of a medical professional! We remain best of friends to this day. I clearly remember looking through a gaping crack in the floor beside the hand-crank printing machine. There was a small creek gurgling just below the floorboards. Such was the construction of many of the early buildings in Homer. Ben Walters kept a loaded pistol in a desk drawer next to his hip flask in the other half of the building, which later became the Homer icon, Cups Restaurant.

Soon, a flickering candle of youthful optimism lured me to pursue the risky money that I dreamed might come from winter king crabbing. I was ready to test my mettle and prove my worth in an environment that has only recently become known as the most dangerous occupation in America.

Going after a job as a deckhand on an old boat of questionable seaworthiness like the *Miss Charlotte* was an uncertain first step on a slippery ladder toward the possibility of a big payday on a bigger, safer boat like the *Shishaldin*. I had a friend aboard, and its home port was Seldovia, thirty miles down the coast. I fantasized that such a berth might give my young wife and me the big grubstake we needed so badly to get serious about building the wilderness lodge of our dreams.

The *Miss Charlotte* had been refitted and pressed into service as a crabber from its former career as a small salmon tender. It was never built for the dangerously top-heavy deck loads of big crab pots. It was a dry boat; its hold was not filled with water to keep the crabs alive. If it had been a wet boat with a fish hold full of water, that would have lowered its center of gravity and made it more stable. Our rolling and pitching was worse still with ice in the rigging far above that sacred center of gravity. I had heard one old-time fisherman quip, "For God's sake, we used to work hard to keep the water out of our boats, now we fill them up with water." Wet boats allowed the crabbers to stay on the grounds longer. We returned to the harbor every night to deliver the crabs alive, even though they might be in the dry tank for hours before delivery. When we dumped the pots on deck we hurried to get the legal-size crabs out of the wind-driven cold, which would quickly kill them. The cannery rejected any dead crab, so we were doubly challenged with a dry boat.

The Homer Boat Harbor, our home port, was frozen over so solidly that you could drive a truck around on the ice inside the harbor. A small steel tug, the *San Carlos*, had been hired by Superintendent Bill Miller at the cannery to break a path through the ice from the harbor entrance to the cannery dock for the king crabbers. Even a careful tug operator pushing ice inside a harbor is looking for trouble. The pushing force of a thick sheet of ice readily communicates horizontally over a great distance. The *Dan 36*, an inlet drift boat, was in its winter mooring and far from the powerful little tug, but one day the pushing ice crushed its stern and it sunk.

The smaller wooden-hulled boats, marginally equipped like ours, left the harbor at daybreak for fishing and steamed back in behind the tug just before dark on those short, hard winter days. This was no place for small wooden-hulled boats. There was a single orange life ring hanging on the front of the wheelhouse but no survival suits and no life jackets to hedge against the forty-degree water. No computers, no cell phones, no pocket-size GPS (global positioning systems), and no coast guard rescue helicopters to increase the sailor's chances in that unforgiving environment. Survival time in that icy water was measured in minutes. No one talked about what to do if you went overboard. My friend Grady slipped on an icy ladder going down to his boat, hit his head, and, unseen, drowned in minutes. The nearness of death even in the harbor remains ever present in winter king crabbing. It's often the unexpected that kills you.

The three of us aboard the vessel *Miss Charlotte* were so broke that we pushed through the Homer Boat Harbor's thick broken ice at dawn and dusk, again and again. The poor old *Charlotte*'s wooden hull and prop were taking a terrific beating. This was not an auspicious beginning for a young man trying to learn the ropes of commercial fishing. The heavy floes banged and crashed into the hull as they rolled over and slid under the keel. The big-bladed prop hammered terribly on submerged floes the size of pool tables. It rattled our teeth and the dishes in the galley. We felt like a bone being shaken and chewed by a dog. The ironbark chafing guards at the bow stem were broken and splintered. As soon as the hull passed, the momentary void behind us filled in with ice as if to prove that we had never existed. We were nothing if not anxious to leave the Homer Boat Harbor for the ice-free port of Seldovia, thirty miles as the gull flies southwest. A stabbing north wind finally drove us out, and we left a trail of frozen mist off our stern as we steamed down the bay. There was a nasty ice-induced vibration coming from the shaft alley on that day, and none of us liked it one darned bit.

Tucked into a protective bay, and clinging to the side of a steep mountain, Seldovia had been a Sugpiaq Native village for centuries. *Seldovy*, Russian for "herring," had been the czar's trading post a hundred years before the US Civil War. The little onion-domed Russian Orthodox church overlooking the town is more than 150 miles south as the gull flies from Anchorage and little more than halfway from Anchorage to Kodiak. No roads connected the town to the rest of the world. Small bush planes landed on the tiny gravel strip with tall spruces at one end and a salt marsh at the other.

The foc'sl of the old *Charlotte* was never designed for a live-aboard crew. Life was a cold, cramped tedium of wet rubber gloves, unwashed clothes stinking of gurry, and flat food from round cans. There was an almost continual dull thumping of the heavy bait ax on the deck overhead, where forty-pound boxes of frozen herring were chopped into pieces small enough for the bait jars. If there were only one word to describe this wet, cramped, and cold world of ours, it would be *miserable*. During those starry nights the wind howled in the rigging and the northern lights danced overhead in arcing rainbows of color. The surge of the sea even in the harbor told of the driving storms just beyond the breakwater.

Bumping alongside the dock at the end of the day, we traded the giant "sea spiders" (as we called the king crabs) for fish tickets that we would exchange later for cash. Some of the really big crabs had a carapace as broad as your chest and a leg that would reach over five feet. From time to time, the pots came aboard so stuffed with crab that you couldn't shove another one in through the entrance tunnel with your boot. Grub and bait were lowered a full thirty feet over the side of the cannery dock when the tide was low. Rusty old cable winches creaked and groaned with the loads.

The winter weather in the open inlet just outside the Seldovia breakwater was often too severe for our little thirty-eight footer. We were stuck in the harbor while the larger vessels came and went. Day after day of cold and ice, wind and storm left us trapped, bored, broke, and restless. The Linwood Bar was no safe harbor either — drinks cost money. I had no problem with that temptation, however; my wallet was as empty as my pocket.

We crabbed as weather permitted between the ferocious gales of Lower Cook Inlet and Kamishak Bay. Winter storms, freezing salt spray, and tidal surges of up to thirty feet of vertical change in six hours set up conditions that sent many a good fellow to a watery grave. Temperatures dropped and waves crashed over the bow and spray flew like a death curse across the decks. Much of the water froze where it was flung. Shrouds — support cables the size of your finger — grew to the diameter of your leg. It was possible that in short order the vessel would build so much ice on deck and in the rigging that the top-heavy boat could roll over. Good catches worth a small fortune could be had around Augustine Volcano, where winds greater than one hundred knots were not unusual during those dark winter months. We fished those dangerous places, glaring deliberately back into the eyes of the reaper. We beat ice off cables and gunwales on surging, wave-washed decks until we staggered from exhaustion. Farther south, down the Shelikof Strait, my brother, skipper of the *Margaret Lynn*, listened helplessly on the marine radio as his partner fishing boat rolled over in a winter gale. All hands were lost.

I was so proud of our meager toehold in China Poot Bay that I could have burst. I wonder now what our few visitors must have thought of the little one-room log cabin and the guest room I had added on. It wasn't much, that was for sure. I guess those folks must have smiled indulgently, reluctant to tell this young fellow that he was crowing like rooster about nothing much at all.

Other than Point Pogibshi — also called Dangerous Cape — it could be said that the most dangerous place in Kachemak Bay is crossing the bar at the entrance to China Poot Bay when a big run-out tide is muscling into the maw of a southwest blow. When we were new to the bay and learning the ropes, I had a deep and abiding fear of that place.

Apologizing to John Masefield as I did, I wrote this poem after one particularly perilous crossing of the China Poot Bay bar in our small wooden dory. Diane was sitting amidships on the floor to keep the center of gravity low. She was hunkered down under an old-style canvas tarp with the brass corner grommets and a six-fathom shot of sisal sewn into the edging. The spray was flying like buckshot and striking me so forcefully in the face that in spite of having my Helly Hansens cinched down so tight my eyes were bugging out, spray had found its way from my throat to my underwear in an icy trickle. I had seen truck-size boulders beside the channel at low tide and feared that when we dropped into the deep troughs, one of them might just come up through the bottom. I was trying to quarter the really big whitecapped rollers as they arched up menacingly over the submerged bar. I made eye contact with Diane and could feel her deep-seated fear. She knew her life was in my hands, just as both of us knew that our lives were in hands bigger than our own. Salt water was dripping off the eyelashes of those bright-blue eyes, infant

Shannon was asleep in her arms, and as terrifying as the situation was, there was a timeless beauty to it.

We have known her many moods,
We have seen her varied faces.
We have walked her driftwood beaches,
We have shared her lonely places.

Working off a windward shore,
Fearing a water grave,
Running with the flood tide,
Laughing with the wave.

She has battered us and bruised us,
When rest we sorely craved,
She has whispered low in sea caves,
For long we've been her slave.

Her thunder on the beaches,
When the sea winds blow,
Her many moods and faces,
We've come to love and know.

And when our span has ended,
And the time has come to go,
Then she will take us home again,
That's all we need to know.

Michael and the family Labrador retriever sit on the porch of their sod-roofed, cedar-lined sauna at Kachemak Bay Wilderness Lodge. Photo © Boyd Norton.

CHAPTER 3

The First Few Years

Life in China Poot Bay

"Those who live surrounded by beauty
become eloquent."

—*Jalaluddin Rumi*

Diane and I had enormous appreciation and respect for the old-timers, those pioneers who paved the way for more recent arrivals like us. It took real grit to come into the country when the roads were little more than trails and homesteaders were scattered far and wide. Roads and schools had been built, and there was an organized community in place when, as a bachelor, I first fell in love with the bay in 1966. In the late 1960s and only seven years after territorial status changed to statehood, I convinced my new bride to move to China Poot Bay. There were already two generations, more or less, of non-Native people scattered across the land: those who came before World War II and those who came after. Our generation was a third group, the Vietnam-era

generation. Without exception, those wonderful old-timers were help-
ful and supportive to us. Their enthusiasm, good spirits, and can-do
attitudes provided steady and invaluable role models for us. Grandma
Wally showed Diane how to make that perfect moose gravy "just right,"
and Sam Pratt, whose name is honored by Homer's museum, showed
me his collection of objects and some of his painting techniques. We
were so broke that Dick and Lynn Inglima gave us credit for groceries,
and Dick and Brantley Edens gave us credit for a drum of fuel oil for
our tiny one-room log cabin. Hugh Watson generously shared valuable
information about hunting and trapping. These pioneers and many
like them whom we came to know and love took time to teach us, to
help us, and thus encouraged us to help others when our turn came.
This was and is the essence of new people coming into the country:
old-timers helping the newcomers. We always felt fortunate to be able
to give back to the community, and in doing so, to honor those tough
and resourceful first non-native pioneers of Alaska. We had and have
the opportunity and privilege of being hospitable and gracious to
visitors — both the newer arrivals and those less fortunate than us.

One day not long after our arrival in China Poot Bay, I found
myself crouched on a narrow stone ledge at the margin of what we
soon came to call "Home Cove." The tide was nearly high and still
rising. The water was so still that I could kneel at its very edge next
to deep water and not be splashed by waves. I tried to ignore the pain
in my knees from the sharp-edged rocks as I leaned forward to peer
down into the sea.

I had tied the neck and innards from a duck I had shot the day
before to a string and added a small rock for a weight. I was tossing it out
into the bay and slowly retrieving it to see what manner of crab or fish
might be drawn by the fresh blood. I had no fishhook to snare whatever

might be down there, so my mission was more about gathering information than putting food on the table. A newcomer in a hard land needs to collect and file away any useful information having to do with living off the land. I had seen this place exposed when the tide had dropped more than twenty vertical feet the day before and knew that the shore pitched steeply into a still-deeper channel — and who knew what lived in those depths? There was an inner child at work there, full of the wonder that comes with peering curiously into deep water. I knew that the people who had lived on these shores for thousands of years had been prosperous in large measure because of their utilization of the food that could be had at the shoreline. The simple food supplies we had brought from town were limited by our near-empty pocketbooks. The wooden boxes of food we had hauled up the beach from the old boat that had dumped us on the shore consisted of staples with no frills. I had hopes that we could supplement those simple supplies with what I could catch, shoot, or trap. It was clear that in order to survive and prosper in this remote Alaskan place, we would need to learn to live by our wits. Any trip to the shoreline was an opportunity to gather food or to learn, without a two-legged teacher, how to harvest whatever was available.

The shoreline concavity that we called Home Cove wrapped around me like a living blanket of stone that caught and amplified the sounds and textures of the place. There was nothing to which you might have given individual names, like a gull's cry or raven's croak. There was, however, an elemental sound and a residential presence — as if you could sense the breath of a person standing next to you.

I retrieved the bait slowly; I could feel it bumping along the bottom. I was as expectant and excited as a child at Christmas. As it neared the surface, I caught sight of the wiggling intestines a few feet below in the clear water and was disappointed to see that they hadn't been touched. Suddenly, the wide eyes of a harbor seal appeared out of the depths. In shock and surprise, the seal's eyes locked with my own for an instant.

In that moment we flashed deep into one another's brains, our noses less than a meter apart. I saw the details of eyelashes, whiskers, and a curious integrity. It realized the enormity of its mistake to have come so close to one so dangerous and virtually exploded in spray. The flippers and tail of the muscular one-hundred-pound pinniped responded with every fiber of its being as it answered the command from the bridge: "Full speed astern." Before I could react, a gallon of icy water flew directly into my face. This was no time for retreat to the warmth of the log cabin, Diane's woodstove, and dry clothes; this was a challenge.

Baited string in hand, I cast to the sea again, ignoring my discomfort. A tiny trickle of salt water weaseled its way down my chest toward my navel and sent a shiver up my spine. Here was the determined hunter-gatherer of old, at work in the age-old process of learning how to hunt. Every meal brought home to a family in a situation like this was a small triumph of trial and error. Hundreds of attempts to secure food were needed for the one success.

There it was again, that tug on my line. Was that something pulling on the bait, or was it caught on an underwater snag? No, there was a movement, a small struggle at the other end — there was something there. I tried to be gentle lest it release its hold before I could see it. Slowly, slowly I coaxed the resistant creature closer and closer. I peered into the green opaqueness, and to my great delight, a large Dungeness crab slowly appeared, magnified by the clear water. The determined crustacean had been invited and was about to accept that invitation to dinner at our house.

As simple as it was, this process was an important juncture for me because I took it as a clear communication from the place directly to me. It said there was a willingness and ability of this site and its creatures to sustain us. I stopped the tug-of-war for a moment to have a better look. My nose was almost touching the water. I could have given Pavlov's dog a demonstration in salivating. It was a beautiful specimen, my first. I could already imagine the red color of its carapace as it cooked in the pot. I imagined the sound of the cracking shell as I cleaned it. I could smell the delicious odor and savor the taste. Hope scuttled forward to combine with strategy. If I let him chew on that meal for a while longer, maybe, just maybe, I could coax him within my reach. I pulled off my shirt, placed it over the jagged rocks in front of me, and lay down with my face nearly in the water. Never look your prey — any prey — in the eyes, I reminded myself. I slowly eased my arm full length up to my shoulder into the ice water hoping that as I eased him closer, he wouldn't let go. Closer, closer, my arm already ached with the cold. He was strong; I was cunning. He didn't want to be pulled upward, for he knew that there was danger in the brighter light near the surface. His eight strong, calcareous limbs resisted my pressure as he tried to hold on to the bottom. My numb fingers tiptoed up behind and tossed him up on the rocks, letting go quickly enough to avoid the snapping pincers. I placed him upside down beside me and began the process again.

As strangers in a strange land, we began with tiny steps forward to feel an integral part of creation. We began to hold a deeper appreciation for the original Native inhabitants, and we reached for a partnership with the life around us rather than detachment from it by virtue of our intellect or reasoning. Now, four decades later, the attachment deepens, and, like the surrounding spruces, each day our roots hold more tenaciously to the rocky soil.

During our first summer in China Poot Bay we rented out the one-room cabin where we were living for $350 for the entire season, and we moved into the trapper's cabin, exactly eight feet by ten feet, with my thirteen-year-old brother who came to spend the summer. The three of us couldn't be inside together unless we were horizontal at bedtime. We cooked on a small camping stove and washed dishes on the front steps. A few years later I climbed up on the roof of the lodge, fired up a chain saw, and cut a hole right through the roof to provide some light into the little sleeping cubby that is accessed by the steps leading up the front of the fireplace. We called these events "progress."

When we were learning to be Alaskans, trapping was as much a part of life as computers are today. It was simply the pattern in the fabric's woof and warp that we pulled over us. For thousands of years the first Alaskans, the Beringial migrants out of Siberia, had used all manner of ingenious traps for animals as big as bears, as small as songbirds. Our land and cabin had been used for trapping; the original cabins at what would one day become our mountain lake camp were trapline cabins. Likewise, when we went farther west into Kamishak Bay to build a brown bear photography camp, we were following in the footsteps of trappers. Anyone who got off the highway system encountered the fascinating world of trappers and trapline cabins. In any room full of old-timers, there would be several who were, or had been, trappers. There was a certain indescribable patina to these people, and we stood in awe of them. Chuckling, a friend of mine said, "Those ole' timers had bark on 'em."

No one has better described trapping and life on a trapline than poet-homesteader John Haines in his book *The Stars, the Snow, the Fire*. In today's world, leg-hold traps and snares are thought to be evil things used by bad people, but like so many other disputed issues, one ought to "walk in the other person's mukluks" before passing judgment on their thoughts and actions. The experiences that are seen and felt and learned on a trapline simply cannot be replicated by any other means. I never lived that life fully, but I lived some of it. There was a time when my optimism about a life with Diane wavered because I could not see her living a remote, hardscrabble, and even dangerous trapper's life, whereas I wanted it very much. I knew that in spite of her rawhide toughness, a trapline cabin was no place for a smart, gentle, and compassionate woman like my pretty blonde darling.

Still etched on my mind is the image of my sweet little gal standing in the Homer Post Office that winter in blue jeans. We had come across the bay in our open boat in winter for supplies and to check the mail. Her beautiful blonde hair was tied in pigtails. Boy, she was cute! She reached for the little brass knob on the frosted glass door of our mailbox and turned it left 9, right 5, left 6. She reached inside, removed an envelope, and tore off the end. As she removed the letter, something fell out. She bent to pick up a check. I watched her study it in surprise. I saw tears suddenly well up in her eyes and streak down her cheeks as she read the letter. It was a $200 deposit from a family from New York who wanted to join us in our newly minted enterprise for a week in June. Until that moment we had been dead broke. We were so discouraged with our future prospects that the sudden good fortune provoked her salty gratitude.

In a fit of extravagance on that trip to town, I bought a rose-colored dress in the front window of Uminski's cramped little pioneer store for her midwinter birthday. She had no need for a dress like that living in a remote log cabin, but I bought it anyway. That night she put on a little dab of "foo foo" as we called her perfume, and she modeled the dress for me; Coco Chanel move over, ooh la la! I gave her my best French wolf whistle. That night as I nuzzled into her neck, drawing in that woman smell, she told me how much she loved the new dress, and I told her how much I admired and respected her for being willing to join me on the little adventures we had shared and the great big one in front of us.

When we first arrived in China Poot Bay and were greeted by those cold, dark months of isolation, it was Mrs. Allen's woodstove and the bread that Diane produced with its kind cooperation that played a pivotal role in keeping us afloat. The work was hard and long and sometimes dangerous, but the bread at breakfast with eggs from our hens, the bread with moose meat on the midday sandwiches, the bread at night with honey and blueberry jam for dessert literally was the staff of life. The bread offered us the credentials that we still lacked as newcomers. We were still those strangers in a strange land no matter how much we felt at home, no matter how much we loved the place. The fact was that the place hadn't yet fully accepted us; we had yet to earn the love and respect of the place where we knew we would spend the rest of our days. The bread, however, built a bridge between us. The smells of the baking bread mingled with the smell of the tide and the spruce forest. The smoke rising from the woodstove danced in lazy spirals with the descendant breezes from the Valley of Four Glaciers. All of our senses rejoiced in the sound of the crackling wood fire mingled with the raucous call of the Steller's jay on the windowsill, the sight of the brown crust next to the golden amber of the logs. The feel of the hot loaf in our hands matched the round bottoms of our children, the smell of a fresh-cut slice intermingled with the smell of Diane's hair as I bent to kiss her in appreciation. Nothing ever matched the sensation of the crunching crust mingled with the highbush cranberry jelly. The bread from the woodstove bore associations that I would carry with me for the rest of my life.

That thin wisp of smoke rising in blue-gray curves from the chimney of our little log cabin on the shore in the Alaskan wilderness could have inspired Alaskan artists Sidney Laurence and Eustice Ziegler, to name a few, to capture this image magnificently in oils. This vision has launched a thousand dreams. Add a backdrop of snow-covered peaks and an eagle soaring above and some bears on the neighboring estuary. Now just for good measure, throw in a sea otter coasting by on the rising tide and a hundred seals asleep on a sandbar. There you have a description of the place where we dropped anchor in a cove with good holding ground.

This inspiring backdrop encouraged a vision of what was possible on the property when we first arrived. It came to pass that we would disassemble, relocate, rebuild, and make better-than-new more than one log trapper's cabin, as we found practical and handsome ways of incorporating them into our emerging business. They gave us and our place on the estuary a sense of history and offered insights into the lives of our predecessors. In the same way, taking our guests to see the ancient pit-house sites of the first Native people nearby helped them look respectfully into the distant past.

With careful attention to detail, we relocated and resurrected one of the cabins to become our beautiful sod-roofed, cedar-lined sauna with stained glass and a picture window looking into a fern-and-rock grotto. There were once many hundreds of log cabins like it all across Alaska, but most have simply melted back into the earth under the weight of time, snow, moss, and lichen. A few of these lovely structures still stand, and usually this is at the hands of a person who wants to honor the old-timers and their old ways. Those forgotten cabins in quiet valleys and on sunny knolls may be largely gone from sight, but we have helped a few of them live on as elegantly as a trumpeter swan gliding in the mist rising from a mountain lake.

The role of our log-cabin sauna in the lives of those associated with the lodge is as central as the bread from Diane's woodstove. At the end of a day, the cares of the world and aching muscles evaporated into the smoke and steam rising from the chimney. New dreams were spawned here too, and this building played a vital part in what happened around it in the ensuing forty years.

Surely, any structure as utilitarian and romantic as a small sod-roofed log-cabin sauna must have a literature, a tradition, and a culture as vast and diverse as one might imagine. As far back in history as you might choose to look, people have sought revitalization through ritual bathing; they have always had special places and traditions for this rite. In the boreal lands, log-cabin saunas were the common denominator among all of us who lived beyond road's end. They were the preferred means of bathing when there was no running water or electricity. This tradition is alive and flourishing today in northern European countries, Alaska, Canada, and Russia.

When distance has lost its allure, when danger has been bravely faced, when the ardor for adventure has cooled, frontier people longed for the rest and revitalization offered by the generously steaming mists of a sauna. We faced threats and challenges both big and small, real and imagined, and saw friends battered, bruised, and even killed by the elements surrounding us. We too would have to pay with a pound of flesh, for such was the cost of living in a wild place as grand and glorious as this.

In the midst of the uncertainty about our chances for success in this spectacular place, Diane beamed; she radiated in her last trimester like all mothers in good health. As I've grown older, I've come to appreciate even more the loveliness of a pregnant woman. Something very deep within the genes of some men stimulates a protective and compassionate response from our brains and hearts when we're near a pregnant woman. Even the most calloused of men are softened. Put your arm around such a woman, offer a shoulder as she navigates stairs, give her a foot-and-calf massage, and just sit quietly with her for a while. Something quite beyond the ordinary will occur among the three of you; such moments are precious and will not come again. Yes, of course the child feels this and develops accordingly. Jane Goodall once observed to me that "when we don't communicate with words, we are open to our primitive selves and then open to mystical experiences." What experience could be more magical than bringing a child into this wild place?

Diane's belly was swelling like a bean pod as winter tightened its grip on the estuary. There was smoke rising from the chimney of our little log cabin on its shore. Over the years, the cabin has been added to with one addition and another, and, like adding pages to a book, its warehouse of stories multiplied as we expanded. The cabin was by now blended so seamlessly into the surrounding spruces, cliffs, coves, sandy beaches, and sunsets that we and it became one with the natural world. We were there to stay. I wanted to believe that we were beginning to evolve into what the first Europeans said about the men who paddled out to greet them in their kayaks: "It was difficult to see where the men left off and where the kayak began, so seamless was the connection between the two."

The time came when Diane awakened me in the middle of the night: "Honey, we have to go to the hospital, I'm going to have the baby." Lucky for us, the water was flat calm and we had a full moon to light the way across the miles of open ocean. The boat was out there in the moonlight, tugging on its mooring, tank full of gas, and fit as a fiddle.

The geese attained a place in our family history on the night that our son, Morgan, was born. When we emerged from the little bedroom rubbing our sleepy eyes and passing through that portal, we were greeted by a scene that has never repeated itself. On the living room floor in front of the dying embers on the hearth was a liquid pool of brilliant moonlight. We stood for a moment, apprehensive about stepping into this apparition. Then we did, and stood transfixed by some powerful magnetism that seemed to weld our feet to the floor. There was immediacy about my young wife's condition, but for that moment in time and space, everything stood still; there was a calmness and we felt such unity — it was as if we had waited every day of our lives for that moment. The moon had overpowered the stars and was strong enough to create our own definitive shadows there on the floor of the log cabin.

The husband was waxing poetic about the stunning beauty of the setting while the wife was thinking, "Okay, buddy, let's get the show on the road or I'm going to have the baby right here on the living room floor."

We had taken some preliminary precautions in the event that this moment might come in the midst of one of the great oceanic storms that pounded our coast at this time of year. A few days before, in preparation for this moment, we had taken Shannon by boat to stay with our neighbors John and Inez. Inez served us the best powder biscuits in the world. We have often mused that Morgan stirred when he felt the biscuits coming down, and they gave him the vitality to grow into the fine, strapping fellow he is today.

Stepping up to the big south-facing window, we could see that there was not a breath of wind. The moonlight lay in a broad path that angled out of the bay toward the volcano — the same direction we would take on the boat trip to town. In that path, our white geese were leaving silvered ripples in their wakes. This was clearly a benediction, a blessing from the land and sea and from the ancient ones. We were standing on the threshold of the birth of a child who had been conceived here, the first child since those original Native inhabitants of so long ago.

I helped Diane across the broken boulders to the shore, where she stood patiently as I popped into the kayak and paddled quickly out to the boat on the mooring. It sputtered to life; I motored in to the shore as she waded out to the boat. We motored out of China Poot Bay and toward the lights of the distant harbor.

Halfway across the bay and right in our line of travel I spied a beautiful log, straight as a string, a terrific log, a world-class log! I would have to dodge to avoid hitting it. Now, perfect logs are hard to come by, and a fellow who is trying to build cabins has great need of such materials. A cabin builder, if he is able to put his logs in water, can tell a great deal about their weight, balance, and trueness based on how they float. Any irregularity reveals itself. I could tell even in the dark that this was a real beauty, about forty feet long with hardly any taper.

"This will just take a moment, honey." I tied it alongside with a quick running-line hitch, and we pulled it into the harbor alongside us. Diane gave birth a few hours later. Today that fine-sanded, amber-colored, oiled, and polished log supports the roof in our lodge dining room. With the retelling Diane smiles benevolently, knowing that for a moment I thought a log was more important than getting to the hospital on time for the birth of our son, Morgan.

In China Poot Bay, I continued my guiding every fall, with most of the clients coming from Europe — and what a dazzling array of clients they were. The baron and baroness of Franconia, the duke of Alba and rightful king of Spain, an ambassador, heads of major European banks, and members of the British Parliament. Goodness knows, it was amazing. *The New York Times* did a Sunday Travel Feature on us, and the list kept growing. How in the blinkin' heck did these people find us? We just happened to be in the right place at the right time with the right tools in our toolbox. They did more than I could describe to make us feel valid when we so badly wanted that verification. Of course, the income from their visits was important, but there was something deeply personal about those particular associations that gave us the feeling that we were in the right place. We began to think that we might make it after all.

I took the state's oral and written exams and was awarded master guide license no. 51 to become one of the youngest guides so recognized. This progression in licensing seemed to be part of the natural flow of events in a career field where there was no recognized PhD program leading to the standard of accomplishment to which I aspired. Commercial pilot's license, US Coast Guard captain's license — these were pieces of paper to put on a wall or list on a CV. But where was there a description of what I should know in order to become the best I could be at what I wanted to do? I studied and read and did my own field trips, focusing on just about every facet of natural history I came in contact with. Ever the smart one, Diane grew leaps and bounds when we lived in town one winter after Morgan was born to put up a grubstake to buy supplies with real jobs. We both became emergency medical technicians, but she went on to teach EMTs, and we lived with the pager that called her out in the middle of the night to respond to plane and car crashes and other fatalities.

There was even an opportunity to study lepidoptery, as there were several conspicuous species of butterflies and moths in our area. Cartography and myrmecology both seemed relevant to a person wanting to know a great deal about the natural world. I joined the Homer Society of Natural History and was lucky to have been acting chair when the paperwork arrived, addressed to me of all people, giving us national accreditation. The dedicated work of volunteers and directors gave a scientific resource base to the little town, providing many of us with the means to learn in ways that would not otherwise have been possible. Our growing skill and information set helped Diane and me realize that the kind of clients we wanted were actively engaged in lifelong learning.

We set about replicating the kind of experiences that visitors might find and that we had seen in Switzerland and Austria, where people came to hike and find pleasure in mountain scenery. We created trails that followed the edges of coastal cliffs for miles. Instead of shooting a bear or moose with a gun, we shot bears with cameras. Instead of focusing on a bigger rainbow trout or salmon, our clients delighted in watching sea otters or a harlequin duck. We invested in canoes and kayaks when ours were the only ones in the region, and we became fair interpretive naturalists in the marine ecosystem. With a teeming seabird rookery just a mile from the living room and a delightful array of passerine and other land birds, we became a birding destination too.

The nonconsumptive philosophy we sought to incorporate into our business was nothing new, but it was not being emphasized in Alaska when we began. We had much in common with Celia Hunter and Ginny Wood at Camp Denali near Denali, so we invited them to visit, cementing a lifelong association. They too had taken on a big challenge in a remote place. Like me, they were both pilots, and together we pioneered ecotourism in Alaska before the word was in common usage. Celia would soon head The Wilderness Society, and years later I would sit at the board table of The Nature Conservancy between Celia and Governor Wally Hickel, as Governor Jay Hammond looked on amusedly at those two noteworthy figures who were of such different political and environmental persuasion. When years later the Alaska Wilderness Recreational and Tourism Association gave us its annual award for leadership in our career field, I was reminded of the times we shared before the word *ecotourism* entered the vocabulary.

They called it China "Pot" in those days probably because China "Poot" made no sense to anyone. We looked it up on the maps and charts and started calling it for its real name and its namesake, Henry "China" Poot; local people soon followed suit. Janet Klein, noted resident author, avocational anthropologist, archaeologist, and lifelong friend, was a fellow team member for many years at the lodge. In the Seldovia archives, she found records of Henry's life. Like me, he spent time in Chenik Lagoon in Kamishak Bay beyond Augustine Volcano. Bear Cove pioneer Ruth Newman once told me, "China Poot used to have a saiwash back in the woods behind your lodge." Indeed, there are several caves where a tarp draped across the entrance with a fire inside would have provided acceptable shelter during Henry's hardscrabble days.

Perhaps most of all we owed thanks to the wilderness — that great living, beating, open heart of the natural world that has battered and bruised us, nurtured and sustained us, and time and again left us full of wonder. We are, our children are, our home and business are, the product of every alder leaf, a tribute to every salmon and clam, a song of praise to every insect and whitewater stream. Collectively they offered a blue-green petri dish where our little experiment could prosper and grow.

As commercial fishermen, we "highlined" Cook Inlet as set netters in 1976. In fisherman's parlance, this means that we caught more salmon than other set netters. With those funds we bought a sawmill, which provided us with the resources to build the summerhouse that today houses the bathrooms and laundry and a little room upstairs. So up those stairs we went and lived for more years, until we built our own cabin on the top of a cliff overlooking the estuary.

We felt like we were making pretty good progress with the lodge as fall rolled into winter. That is to say, living in China Poot Bay and actually building a business there seemed like it might actually work. Before then, there was serious doubt about the wisdom of having burned our bridges by leaving our jobs in Homer. Having made our deliberate move across the bay, it was now a matter of sink or swim. There had been enough paying clients during the previous summer and fall to pay the mortgage and fill our food cache with winter groceries. My clients had taken moose, mountain goat, caribou, Sitka black-tailed deer, and black bear. Each provided an excellent store of protein. The vegetable garden had been bountiful, and there were laying hens on the south-facing hillside above the greenhouse. The bounty and our harvest were meager by some standards, but we thought we were rich. In the hope of soothing Diane's parents' concerns about our well-being, I reminded them that we had "a low standard of living but a high quality of life."

We had weathered the storm of baby Shannon's emergency trip to the hospital and the expensive surgery that went with it. We had survived my own trip to the hospital when I tried to load clients into our commercial fishing boat in the surf in front of the lodge. I had been pinned between the hull and a boulder during a frightening storm, and my ankle was crushed. All of this happened at the end of the summer, just as my big-game-hunting clients were to arrive, bringing money that was supposed to get us through the winter with food and more building supplies.

I had been asked by the editor to write an article for an *Alaskan* magazine about the ethics of hunting and fishing in Alaska. I observed that if the sportsperson would measure the success or failure of a trip not by the number and size of trophies but rather by the overall sense of well-being that the experience generated, then the percentages of "success" on any trip became very high indeed. I wrote that it is simply not possible for all trips to be successful, if the definition of that success is too narrow. Enjoying the trip as well as the destination is an ancient saying, and I tried to weave it into the Alaskan scene.

One of our boats was ripped from its mooring by a winter storm, never to be seen again. While we were guiding for a German client, my assistant guide, Mitchell, whom I deeply admired and loved like a younger brother, drowned while we were on a moose hunt at Tustumena Lake. There were more accidental deaths whose long shadows lingered painfully, and other tragedies were about to cast their dark and heavy pall over us. In the midst of the grief and heartaches, however, there were sunsets and sunrises, the rooster turning us out of bed in the morning, big potatoes popping out of the garden, and a subsistence net jumping with silver salmon. Mingled among the tears, depressions, and black despair, there were northern lights, the laughter of children, the first glug-glug of the perking coffeepot on the woodstove, the thunder on the outside beaches when the gale winds blew, and joys aplenty.

We had proved to ourselves, if to no one else, that we had the "right stuff": perseverance, passion, and persistence. In those early days, though, it was sometimes a struggle to put on a brave face. If we had not been optimistic and confident, we wouldn't have made it across the dangerous bar guarding our little bay. We understood the importance of maintaining a sense of humility in the face of the giant forces around us. We had drawn a line in the dirt that said we would remain there or be consumed by the overpowering might of our surroundings.

Diane and I were busy at the kitchen table reviewing the winter bills and considering how we would spend the deposits paid in advance by clients who would come for the summer. We had greater needs than we had income, but we felt good about where we were, what we were doing, and where we were going.

Out of the blue, Shannon was screaming as she burst through the door. "Morgan fell in the water!"

Dear God, no — a parent's worst nightmare. I looked out the window and was struck by a bolt of lightning. Morgan was floating face down in the ocean and was just about to be carried away by the ebbing tide.

"Please, God, oh dear God, please… *Nooooooooooo!*" We had glimpsed an image that reached to the bottoms of our souls.

I exploded out of the house, Diane hard on my heels. Down the rocky shore we flew, scattering rocks as we went. I charged into the icy water, so cold it burned like fire. I was waist deep before I reached him. When I picked up the little guy, who was only three, who had been my son, it was clear to me that he was dead. His eyes were open and staring; there was a piece of green seaweed in one. His skin was pale, and he was as stiff as a board. I was completely out of my head. I staggered up the beach like a drunk, clutching my tiny son to my chest, sobbing convulsively. I had the presence of mind to drop to my knees on the rocks and place him on his side on the wooden ramp at the edge of the shore. I began with little puffs of mouth-to-mouth resuscitation as I had been taught. Kneeling on the broken rocks, I slowly coaxed life back into him. He coughed and began to gasp. Scalding tears coursed in rivulets down our cheeks as we too gasped for breath. Diane and I dog-trotted up the rocky shore to the house knowing that he was not yet out of the danger of hypothermia. She stripped off Morgan's yellow snowsuit and her own sweater, and I wrapped a blanket around the two of them so he could absorb her body heat. He was groggy, in and out of consciousness. We knew he could still die as easily as a bird sheds a feather.

We called our friends in Homer on the citizens band radio, knowing that all up and down the bay our emergency would be coming over speakers in living rooms and kitchens, offices, boats, and businesses. The doctor relayed his advice over the phone to our friends, and they in turn repeated what he said over the radio. We were told to keep him warm, not to let him go to sleep, and to get him to the hospital as fast as we possibly could. When Morgan arrived at the hospital, he still had so much water in his lungs that the doctor was surprised he survived.

As children are likely to do, he recovered quickly in the days ahead, but each of us remained permanently scarred by the event — none of us more than Morgan. My own father told the story of drowning in a little farm pond in Wisconsin as a child. He recalled going down and resting on the bottom in a white light of contentment. Those of us who have gone to the other side and returned all seem to speak of it more or less in those terms. My brother came within the blink of an eye of drowning. His boat was flipped end over end while crossing the bar on the way to the harbor in a fall storm. He clung to its upturned bow for two hours against all hope of being rescued from the rip that we had come to call "the washing machine." Diane too had felt herself slip into that white space in Homer's hospital while being stitched up from an accident. When she returned, she bolted up from the gurney and hugged the doctor with all her might, thanking him for saving her life.

Nothing but a close encounter with death can make a person feel more vibrantly alive and more appreciative of the next breath. There come moments when the wonderment of the simple continuum of life is itself so magical that it is as if a pixie had sprinkled stardust over everything and everyone. We lived in our remote setting as similar people have, facing similar circumstances the world over and across time.

Horses have a way of returning to their corrals even after they have thrown their rider. So it was that one day Morgan was again playing outside with his sister. We never let him out of sight now without a life preserver in one form or another. There were a succession of CO_2 inflatable devices, life vests, and even an orange life jacket shaped like an ox's yoke. One of the gizmos was a nylon belt strap with an attached grapefruit-size plastic sphere containing an inflatable CO_2 cartridge. If the child fell in, the water activated the compressed air mechanism and *whoosh*.

One day, we heard Morgan's terrified screaming and came running, this time to the back door — no water in sight, thank you. He was standing on the back steps from which he had pointed a stream of pee at too low a trajectory. The aim was not exactly that of an NRA marksman, and the dwindling flow triggered the gizmo. The ensuing inflation scared him half to death while adding a few more gray hairs to his pitiful parents' hair.

When we join hands in a circle at mealtime, or when we are taking leave of one another, we acknowledge a connection between us and our little place beside the estuary that is deeply spiritual. Some of my teachers have instilled in me an ancient way for the mind to take hold and deal with the mysteries of life and death that constantly surround us all. Ancient scriptures coming out of the Fertile Crescent region observe in metaphorical context that even the gods envy human birth, because only within time is there free will and movement. By contrast, eternity, as explained in these teachings, is utter stillness that exists only outside of time, expressed as that point where future meets past. Seen this way, the present is eternity, and by quieting the mind in prayer and meditation — even in a momentary act of thankfulness at meals or before departure — this simple momentary pausing can help us find eternity in the present. Being completely alive right now seems to be what it is all about. If one understands the value of time, the envy of the gods, then one has a chance of understanding the very fabric of the universe and the place of the heart within it. Our little family found that the vision of peace and harmony was accessible in our little cabin beside the sea.

Morgan grew up to be a fine man that any parents would be enormously proud of. In addition to being one of my best friends, I make light of Morgan's role as the family priest/monk. He sets a standard for the rest of us to follow. He would never tell you about his work in the gutters of Calcutta with Mother Teresa's Sisters of Charity or about helping homeless children in Peru. He paid a very heavy physical toll with his own health — malnutrition, parasites, not to mention the trauma of being in a leprosy ward and trying to make people comfortable as they lay dying. In so doing, he set a benchmark for me and anyone who knows him. He showed us all what it means to be compassionate. He prefers that I not refer to him as my teacher, but in so doing, he brings me full circle to my own first teacher, Sensai San in the Zen garden at Ashiya-Gawa.

Shannon, Michael, Diane, and Morgan McBride. Photo © Boyd Norton.

CHAPTER 4

Raising a Family

"Dad, you better tell Don Fell to move his helicopter,
the tide is coming in."

—*Morgan McBride at five years*

Diane is the unsung heroine in the story of China Poot Bay. She was the cement between the rocks, the byssus on the mussels, the yeast in the bread, not to mention the brains in the business. Her lipstick was always close at hand, for this was a woman, a fine woman — one who took fair pride in herself, her home, her husband, and her children. In addition to all that, she made bread that was so good the pope would have traded his slippers for just a slice.

When I wanted to make her laugh, I'd grab my lovely ol' Martin D12 guitar that I got from a friend by trading a warm overcoat, and then I'd croon a song titled "Wooo-Man, I Mean a Wooooooo-Man." The singer praises his woman, describing all the things she can do simultaneously: bake the bread, split the kindling, change the diaper, damper the stove, balance the checkbook, answer the radio phone, change the oil in the generator, wash the shirt, and wipe the counter, all before you can count from one to nine. Yep, that was my little girl, prettier than a rose and tougher than a boot. We would never have gotten to first base let alone hit a home run if it hadn't been for her pluck, determination, multitasking ability, and her unwavering dedication to our dream.

One of the biggest challenges Diane and I faced when the children were small was how to provide a quality education for them when their peers in town began to catch the school bus. We did not take this subject lightly. We had both been read to voluminously as children, and we read to our little urchins starting even before the time they could fully comprehend. We believed that the susurrating sound of our voices would penetrate deep into their brains and hearts. *The Wind in the Willows*, *The Little Prince*, *The Adventures of Tintin*, and hundreds of other books came alive in front of the great fireplace with its curling flames and dancing shadows.

Removed from TV and late-breaking news and rarely seeing a newspaper, we knew that the children were growing up in a way very different from most of their peers. So, we purposefully set about creating experiences so powerful that they would shape their young lives.

Morgan was five years old the year we went to Nunivak Island in the Bering Sea, where he went to school with the Inupiat Eskimo children. There, our Native friend took us on a musk-ox hunt out on the sea ice where our host shot and we helped skin the remarkable animal at twenty degrees below zero. For years, that shaggy hide played a funny-scary role in our annual hosting of the fourth-grade week at the lodge, when the entire class came with parents and teachers. The beach campfire on the last night always had the mysterious appearance out of the darkness of "Ugga Bugga," whose identity was hidden under the great mound of long hair.

Morgan's sixteenth birthday present from me was an eleven-day summer-solstice canoe trip from the headwaters of the Noatak River in the Gates of the Arctic. And years later, the four of us bundled up two folding double kayaks and flew to Ireland, where we paddled from the headwaters of the Shannon River at Shannon's Pot on a rainy, windy, and mud-spattered trip. None of us has ever tired of adventurous and educational travels.

But before we included the spice of international travel in our children's educational soup, we were blessed to share our home with clients from all over the world, as our lodge continued to gain an international reputation and spread its wings. The children were as much a curiosity to our guests as the coins from foreign lands were to them when placed in their eager little palms. Something quite magical happened when Shannon took a visitor up to the henhouse for eggs, or when Morgan led a hike to the lagoon. We heard decades later from clients who still remembered such things. The children grew up with a high level of comfort in interacting with diverse cultures and languages, and it has served them very well. Either of them could be dropped off at night in some strange country, and by dawn they would have laughing friends, food, and a dry place to sleep.

We were always delighted by the opportunity to expose the children to new cultures when we hosted guests from other countries. And were impressed at their retention rates growing up in a place without television or its related distractions. Each guest from abroad brought something for the children, and each left something indelible with them.

We were constantly amazed at the people who walked up the ramp across the rocky beach to the lodge, our home. Many of this fine and serendipitous assemblage added immeasurably to our lives and those of our children by being interesting and interested. There were the two lady guests who one day wore the personae of well-bred British gentility, but now here they were in jeans and hiking boots at a little wilderness lodge. We liked these two women immediately. At times like these, we felt truly privileged having such wonderful clients. It was as if we should have been paying them, not the other way around.

The ladies had just returned from a vigorous hike through the forest and along the Cliffside Trail to the high point overlooking the distant volcano and the bay, and it was time for a cup of English breakfast tea at the lodge. We had been given half a maple shuffleboard table that I refurbished to fit perfectly in front of the big picture window in the kitchen. This was the best place in the universe to sit and watch the birds plunging for fish and sea otters drifting out with the tide while cracking clams on their chests. There was a tanned piece of moose hide tacked to the header log over the window. On it I had written carefully with India ink and a fine nib pen: "Just as the eyes are the windows of the human soul, so these windows are the eyes of god." I really did believe that sitting there on those little barrel stools with your elbows on the polished bird's-eye maple, you could look right into the soul of creation. Once when I came in from outside, Erma Bombeck, a well-known author and humorist, was sitting there with her chin in her hands. She looked up dreamily at me and said, "If someone lived here, why would they ever leave?"

Fire on the ground, then on the hearth, and eventually in the woodstove was the axis of life in human evolution, and, like a musician plays a harp, we played consciously with this setting and the emotions it evoked. The visitors were affected deeply but without *quite* knowing why. There was a certain alchemy — magic, if you will — that emanated from Diane's kitchen, especially when she was baking bread. It was safe and warm in a way that a child feels safe in its mother's or father's arms. It was an incubator of dreams; it nurtured the best in us. It was our anchor to windward in storms; it balanced our inner and the outer wilderness. For years, the children did their homeschooling at the table beside the stove within reach of a stolen slice. Plans were drawn for docks and buildings on napkins and paper plates, puzzles were assembled, books read, stories written, tears shed, paintings painted, crosswords completed, and even chain saws repaired in the days before we could afford to build a workshop. The kitchen table had seen a lot, and the woodstove had witnessed it all.

Diane had just taken her usual batch of four loaves of bread out of the oven. The crusts were as golden brown as a chestnut, and you could tell just by looking that the crust would be crunchy. She tapped the bread pans with the wooden handle of her favorite French kitchen knife, and the loaves tumbled out like playful puppies on top of the still-hot stove. Quick as a wink she picked them up and placed each one on the stone stairs leading up to the sleeping loft, where they would cool slowly. Their resting places were the giant rock ledges protruding from the face of an enormous fireplace that dominated the living room. It was a fulsome sixteen feet across its face and five feet thick. Diane and I slept in the crawl space above the kitchen for many years and accessed it by those same stone ledges.

Searching for a fine cup of tea, the two British ladies entered, bringing with them their quiet sophistication and noble bearing. The loaves smiled to each other knowing that their destiny was at hand. No timid loaves these, they trembled with excitement, waiting to be spread with golden butter and our favorite homemade blueberry jam. Diane greeted the hikers, and they chatted about the cavorting whales they had seen from Moosehead Point at the end of the hiking trail. "Why, yes, Earl Grey would be delightful," they said as my girl poured boiling water over the tea bags in the Alaskan-style mugs made by our potter friends. Diane sliced thick slabs of bread, applied a generous dollop of butter to each, and placed the bread and jam before them with a flourish.

She had work to do: dinner to make for family, staff, and guests. She turned back to the work at hand with accustomed efficiency.

"Oh, my goodness, what is this divine condiment, this minced puree? It seems to contain a bit of rosemary or perhaps mint. Do tell us how you make it, or is it some closely held family secret?"

Diane stopped her work, cocked her head just a little, pursed her lips, and wondered. A puzzled frown creased her forehead. "What the heck are they talking about?" she wondered. "I didn't put any puree on that bread."

Guests were always curious about our sleeping loft in the crawl space above the kitchen. Evidently, someone had climbed up the great stone ledges on the front of the fireplace to take a look through our tiny built-by-elves bedroom door. That someone apparently had been wearing Vibram soles and had stepped in the goose poo on the way into the kitchen. It must have dislodged on the step behind the stove. The wedge had melted into the hot bread, and voilà! Diane stepped up behind one of the women, glanced down at the plate, and before she could catch herself, blurted out, "Oh my God, it's goose shit!"

The ladies were out the door before you could say Jiminy Cricket. Our frugal ways in that era were such that chances are she scraped off that lovely puree, cut off the bite marks, flipped it over and served it to me without a word. Of course I raved about it too, as I always raved about her bread. It was delicious.

There were many things about homeschooling that we liked very much, not the least of which was the fact that it seemed so good, so noble, so designed to build family relationships that we would treasure throughout our lives. This observation, however, requires a studious look at who the "we" were in this curious equation. Shannon was working on her master's degree in hiding; Morgan of course followed his sister's lead with a bachelor's in following. When schooltime came around, they were not to be seen. Disappearing like a magician's trick was, to some degree, funny and endearing, but broken rules followed by discipline was never easy. There were just four of us in the little cabin all winter, and back then we were still reeling from Morgan's near-drowning experience. Diane was really good at the serious undertaking in front of her and worked hard at it, realizing that those early years of schooling would have a great deal to do with how the children approached learning for the rest of their lives. To her great credit, the children avoid television, read voraciously, have large vocabularies, and love learning to this day. Their mom sought and was awarded a university master's degree in elementary education when they were still in school.

I was the team cheerleader, but I didn't formally wear a teacher's hat during those years — although I hope I qualified as one of their educators. My job in the hierarchy was to quote Mark Twain: "Don't let your education get in the way of yer' learnin.'" It remains a weak-kneed excuse to say that I was running a sawmill, driving a bulldozer, working on a seawall, and building cabins. My nickname for myself in that era was Sisyphus, since I had so much in common with the strong, if small-brained character of Greek legend who was condemned to push a large boulder up a hill only to have it roll back down every time he got it near the top. The difference with me was that I didn't get out of the way of the boulder and got smashed for my trouble.

Diane's excellence as a teacher made up for my shortcomings, so it was easy for me to devote most of my time to the foundational elements of a business that would sustain us. This involved what seemed like an endless amount of time engaged in advertising and promoting. Diane's grandfather's turn-of-the-century portable Royal typewriter served me well. Hunt-and-peck was my system, with carbon copies but no correction ribbon or fluid, so each page was full of erasure marks. Onionskin or blue paper was used for lightweight overseas mail. I could be heard at the living room desk, peck-peck-pecking like a hungry and neurotic chicken, early and late.

By now Shannon was a strong and nimble little girl and her brother, Morgan, was quick and resilient. He was nearly able to keep up with his sister, who was always full of beans. The Nubian goats that we brought across the bay to supplement the children's need for fresh milk had been complemented with a Noah's-ark barnyard of animals. We kept the goat milk in a box I built to fit into the spring. A burlap rag over the top wicked water up from the stream and, evaporating, kept it cool. There were raucous white geese, chickens of several varieties with a full-throated rooster, some ducks, and we were never without a black Labrador retriever. With just a half-dozen does and one buck, we had a single butchering of rabbits and put about thirty of them in the freezer. We had electricity by now, having paid the electric company to run the long wire to tie into a trunk line that ran from Seldovia to Halibut Cove. The antique oil-fired plant in Seldovia also ran power to Port Graham and Nanwalek, Native villages down the coast from Seldovia.

When our nets provided them, we traded salmon to neighbors up and down the coast for vegetables and other needed supplies. On a sunny day, the large greenhouse on the bright south-facing hill above the lodge became a piece of tomato-smelling heaven. It was sixteen by thirty feet, with a full-faced facade built from surplus south-facing windows purchased at $6 each. The garden plants and those who tended them had an elevated view of the estuary that was the envy of the world. I elevated the seedbeds to waist height in sturdy bins so Diane wouldn't have to bend over to weed and harvest. We harvested bushels of zucchini, mouth-watering tomatoes, and several kinds of lettuce, chard, and kale. The outside garden produced potatoes, cabbage, carrots, and one year a monster turnip that, with its foliage, was taller than seven-year-old Shannon. The woods were full of berries in season, and a generous crop was never far from the greenhouse. Fall berry picking in competition with the hungry bears was exciting and fulfilling for the children year after year as they sprung from bean sprouts into preteens. When fingers were stained purple with blueberry juice, it was a good time for a little brother to poke his head up out of the bushes like a jack-in-the-box and stick his purple tongue out, scaring his sister. What's more fun than to gross out your sister!

Using their little red wagon, the three of us dug into the hillside not far from the door of the log cabin, and above that hole they helped me build a three-tiered smoker. It was far enough from the building that if it did catch on fire — and it did — it would not threaten the lodge. It was close enough that it was in the center of our activities and near at hand for regular tending of the slow fire, usually by Morgan.

The first story was lined and floored with huge flat rocks that held the heat and prevented the sidewalls from catching on fire. The second story provided room for the rising smoke to cool, and the third story was accessed by climbing the hill behind the smokehouse to access the wire trays. To reach the trays, a heavy door opened downward, where it rested on a post to provide a working platform for loading and unloading the trays. The ritual of placing salted fish, salmon, halibut, and trout was one we embraced. I loved giving Diane an adoring hug and smelling the smoke in her hair and clothes. No elixir from Coco Chanel could have made her more adorable in those homesteading days. We tried everything in that smoker: black and brown bear meat, clams, mussels, wolf-eels, moose, goats, caribou — you name it.

When once-proud wooden commercial fishing boats are rotting on the shore, when children no longer tend smokehouse fires, when people find it easier to buy vegetables from the store than to tend a garden and have given up or forgotten their attachment to the land and sea, something will be lost that will be difficult — if not impossible — to replace. It is not so much the forgetting of how to do a certain thing but the loss of the attitudes and points of view of the people who did them. Having old-timers with old-time skills and knowledge among younger people has throughout human history added to the next generation in ways that are real and palpable but difficult to measure.

From the gravel bar in front of the lodge we had clams and mussels in such profusion that we were continually staggered by the numbers. We ate scores of octopus, until Diane read Jacques Cousteau's book *Octopus and Squid: The Soft Intelligence* to the children and decided that these fascinating and intelligent neighbors were much too clever, too endearing to eat. Years later, on one of my trips to the cannery dock to deliver my pot-shrimp catch, with Shannon driving the boat, I encountered Steve Nathanson just off his boat, the *Altair*. It was a broad-beamed, deep-drafted oceanic sailing schooner that he had converted to a king crabber.

"Hey, Mike, you ought to have the cannery send you out a box of squid. As I was chopping it up for my king crab pot-bait jars, I thought it looked awfully nice. That's calamari, you know. Well, daughter Patty cooked some up below deck, and it was delicious."

I did have the boys send out a case, and we ate it for the rest of the winter. It was a pity to chop up that fine calamari for bait. I imagined Diane writing a letter to her parents, "Dear Mom and Dad, we ate the bait."

In the tide pools in front of the lodge we learned to make "uni" sandwiches using sea urchin eggs and sea lettuce in a nutritious little snack that wasn't likely to make it on the McDonald's menu. Try as I might, I made little progress telling the kids that the yellow-orange egg skeins in the urchins tasted like cantaloupe. Calling them "gonad gulpers" when they ate one didn't help either, although it did provide the opportunity to explain how the hermaphroditic crustacean sexually reproduced. They challenged me to live up to my own declaration that every sea vegetable in sight was edible, exempting the seaweed-looking hydroids below lodge point where the opalescent nudibranchs feasted on them. Red dulse and sugar kelp, macrocystis, green sea lettuce, and fuchus all crossed my lips to the chagrin of the children, who weren't impressed and had the cheek to laugh at their father.

As the children grew older, they increasingly looked to town for all that they were missing as they grew up alone. We often invited children of similar age to stay with us, but it did not satisfy their craving for more social interaction with their peers, and their teen years were within sight. One day, Diane said, "This isn't really working as we had planned, Michael. I think I will rent an apartment in town, put the children in school, and let's see how it goes." This was certainly not what I wanted, but the majority, including the nonvoting subteen members, had been dutifully polled, and I was no longer leading the pack. It was my turn to fall to the floor, roll over on my back, arms and feet flailing in the air, and hope someone would rub my belly and say, "Good dog, good dog." The benevolent dictator who had been educated beyond his intelligence had at last been overthrown in an overnight bloodless coup. I cried, "Oh, please don't make me go to town and get a real job." This one-lined song was getting repetitive.

I was devastated with the idea of having to live in Homer to be with them, abandoning my winter work, and potentially scuttling the ship that carried my vision. My choice was to stay in my solitary cave, do my thing, and they could visit their husband and dad as circumstances and weather allowed, which clearly wouldn't be often given the difficulty and dangers of crossing the bay during those very severe winters. A party could polka dance on the ice in the boat harbor that was frozen for months at a time. I figured I could learn to fly a helicopter, but the chances of buying one looked slim. Given the fact that in moderate weather I had to be with clients in order to pay the mortgage and buy the groceries, and given that all the advancement of the infrastructure had to happen in the winter when there were no clients or income, I was between the proverbial rock and hard place.

It was rough, really rough; there was a steep-troughed beam sea, and we were loaded to the scuppers. I probably made it harder than it really needed to be. I knew what we could accomplish and how to do it, but my partner wasn't sure that it was worth it. She did not share my clarity of vision or single-mindedness of purpose. It was clear what was best for the children. It was clear that growing up without a surrounding peer group was a valid issue. It was clear to me that my wife and children should be happy, and it was my responsibility to do my part to make this happen.

It is a very small number of women who can be content in a remote place for years on end without the company of other women, friends, and all that socializing offers. The fact was, we later learned, Diane suffered from SAD, or seasonal affective disorder, the downstream effect of light deprivation to the optic nerve and its manifestations as depression. In its worst cases, alcoholism, drug abuse, weight gain, lack of interest in life, listlessness, and even suicide can be common associations. In a remote situation, this condition was a recipe for disaster. Little log cabins were never known as brightly lit environments. The long dark winters and small confining space were of course just too difficult for her. Here again, brave trouper that she was, she fought back diligently to keep our little family together, even though we were often separated by the wild and stormy ocean.

All children growing up away from town develop close associations with animals to make up for the absence of other children their own age. One day, my brother and I decided that a huge spruce tree was growing too close to one of our buildings and we agreed that the continued growth of its trunk and root structure would eventually threaten the building's foundation. With the chain sharpened, the air filter cleaned, and the gas and bar oil topped off, we set to work. When the tree fell, out of the upper branches tumbled a pair of flightless young crows.

We untangled the bucket-size nest from the branches, put it in a Blazo box, and returned the juveniles to their ingeniously made home. The parents scolded us loudly from above but soon abandoned the chicks and flew off to feed and raise a new brood.

Now it was the children's challenge to feed the squawking feather balls, and we had no idea what we were getting involved in. These little loudmouths, named Heckle and Jeckle, wanted to be fed all day and all night long and complained loudly whenever we were a few minutes late with their hourly feedings. They grew quickly on a connoisseur's diet of fresh clams and mussels that Morgan and Shannon collected in the front yard, the food that was most available to us and easiest to feed to them. Since they hatched in the spring, they entertained our guests all summer long with their antics. They were nothing if not raucous when time came for hors d'oeuvres with guests on the dock before dinner.

Throughout the years, a long list of prominent magazine writers and photographers, from *National Geographic* to *Men's Journal*, came on assignment to write about and photograph the lodge. Somehow, an image appeared in *Smithsonian Magazine* of me with a crow perched atop my head. It was of course Heckle kindly giving me a head massage with those little talons of his in appreciation for his clam dinner. Another photographer posted a photo of Morgan in the Library of Congress for an exposition titled "America Reads." He was reading to his black Lab pup on the end of the dock with the bow of the wooden Opheim skiff in the foreground. Asked to submit a photo, I contributed one of Diane reading to the children on the couch in front of the fireplace and another of Morgan reading on the little log bridge across the creek I named for him in Chenik Lagoon in Kamishak Bay.

The children loved to help their mom in the kitchen and happened to look out the window to where our black Lab was sleeping in his favorite grassy spot in the sun. Jeckle the crow waited until he was sure the dog was asleep, then he tiptoed over and began pecking at the dog's toenails. The crow chuckled as the dog's feet began to dance in his dreams. Finally the dog woke with a start, and the crow sprang up to a spruce branch with a loud "haw, haw, haw." The dog was soon asleep and the crow began its pesky business all over again. After a few more times, the dog had had enough. This time it pretended to sleep. The too-big-for-its-britches crow sauntered up and the dog lunged with teeth bared. He narrowly missed the crow, who lost a couple of tail feathers for his naughtiness. As teachers, we pointed out to our children the lessons from the natural world that occurred all around them. Good education is good theater.

One day, we saw the corvids dart into and back out of our loft bedroom when the window was left open. These smart birds are collectors of curious objects, so we were suspicious about what they were up to. Sitting at the kitchen-window table doing homework, Shannon looked up and cried out, "The crow just flew by with the crapper in his beak!" Sure enough, their grandfather had sent to the children the dollhouse he had made for Diane when she was little. All of the furniture was there, including a tiny antique toilet named for the plumber of old, Thomas Crapper. Later that year, I climbed the rope ladder up into the crow's nest at high tide to unfurl the flag, and there in the crow's nest sat the little toilet with bottle caps, coins, tinfoil, and other purloined treasures.

We regularly had the privilege of introducing clients to one another at our dining room table. They came as total strangers, and by the end of the week left with a bond set so firmly that some lasted a lifetime. We pondered whom to sit next to whom at the lavish dining room table under the polished log. Working on this helped me transfer the senses of a builder who works with wood and those of a musician who works with sound into creating situations using the opportunities at hand to please the clients. We worked hard to present people with experiences so powerful that their lives would be shaped by encounters with nature and wilderness. Our children were watching. They took great pride in their roles of table setters and guest servers. Their abundant rewards came in the generous approval they received, and this taught them more than we could have hoped about skillful interaction and being of service to others.

There was great reward in this form of creativity, especially when children of clients were involved and we could introduce them to the wonders of a tidepool, the hush of the forest, or the clatter of stones as the surf hurled itself against the shore. Children's faces are open books to be read, as compared to the studied countenances of adults who have learned to be not quite so demonstrative. We were often asked if this age or that was old enough to bring children into our little piece of wilderness. "Any age," we would tell them. "Even prenatally, the child will be affected by the parents' sense of well-being, and who can say that this is not as important as the child's actually remembering any exact experience."

Years later, Morgan would write what became one of my favorite lines from any poem I've ever read. He wrote about the big seasonal tides and how they assist the annual migrations of anadromous fish like salmon to their natal streams. He likened these energy pulsations to the human heartbeat itself, calling them "Pacific upsurge virtues." Living on the wild coast, I began to understand how the salmon energies were like chemical nutrients carried by the blood in the human body to its most remote cells. The salmon and sea-run trout and char leave the sea; they ascend the rivers, the streams, and rivulets until no tiny trickle, however insignificant, fails to recognize the tickling of a fish's belly on its gravels. From there, bears and wolves, wolverines and ermines, coyotes, river otters, and their furry friends carry the flesh and bones of the fish up and away from the watercourses. Hawks and eagles, ravens and owls and their avian kin, all carry fish to distant treetops and hidden mountain crags. The work of flies, worms, and lesser invertebrates participate too in this magical transformation. The feces of all these animals is spread far and wide until literally no blade of grass, no spruce, no birch tree is unaffected by the millions of tons of fish spread like a huge gossamer mesh across the vast stretches of the northern wilderness. This distribution of energetic material is, as my son describes, the surging virtue offered up from the Pacific sea goddess to its brother, the land.

In the blink of an eye, *einen augenblick*, we watched the children grow into fine young people. What we saw, the children saw, what we learned, they learned. Morgan the tyke, struggling to carry a piece of too-heavy firewood to fill the woodbox, was soon filing a chain saw and felling a tree exactly where he wanted it to land. The little girl who led her little brother up the hill to collect tiny bantam eggs hidden under the henhouse was suddenly in high school and running high hurdles on the track team. Distinguishing herself again, we puddled up as she left for Leningrad, having been first a local, then state, then national, and finally an international winner in a youth leadership program.

It was a bright spring day when Bill and Marion decided to fly their Piper Super Cub the hour and a half south to Kachemak Bay from Anchorage. The flight along the shoulder of the Kenai Peninsula wilderness where it skirts up to the snowcapped and glaciated mountains is as pretty a route as you can find in Alaska. The wide inlet lies flat and huge off the starboard wing, with the spruce, birch, and cottonwood forests expanding across tens of thousands of wild acres. Off the right-wing tip was the broad sweep of the Alaska Range across Cook Inlet, undulating south and west from twenty-thousand-foot Denali, Mount Spurr, Redoubt, Iliamna to Augustine Island Volcano, and finally Cape Douglas and Mount Douglas at the beginning of the Aleutian Range. The cape gives way to the Alaska Peninsula and the one-thousand-mile arc of the Aleutians. On a clear day like the one they chose, this hemispheric display tumbled out neck and crop. Over Skilak and Tustumena Lakes, down the fabled Fox River Valley, and over the picturesque estuaries on the east side of Kachemak Bay, they would fly and land on a beach near our fledgling lodge.

They had relayed word of their coming through our answering-service friends in Homer, and that message came to us the evening before through the weird static, squeaks, and squeals of our elementary communication system.

They buzzed the house the next day. Diane had left a few days before, flying to her parents' home in Libertyville, north of The Big Windy. Morgan and Shannon were five and seven. I don't know how old I was, but I was lonely already and missing my sweet little blue-eyed wife who was happy to reunite with her family for a few weeks.

We watched the little bush plane disappear behind the tall spruces to the north, the distinctive chatter of the Lycoming 150 trailing behind. Listening carefully in order to know what to do next, we heard them land on the outside beach. The three of us hurried off into the thick spruces behind the lodge. Shannon was striding like the athletic little woman she was, and Morgan's short legs were just a'churning. Our visitors had just finished tying down the wings of the Cub to driftwood tree trunks above the high-tide line when we arrived with welcoming shouts and laughter. This was the first time we had seen anyone land on this beach, and we were especially excited after a long, isolated winter of homeschooling at the kitchen table, with no visitors and few trips to town.

We hiked together back to the lodge, the guy wearing my hat chattering like a magpie. Since the tide was low, I offered to prepare a clam-and-mussel lunch as soon as we could dig and gather the mollusks. In nothing flat, we had more than we could eat and an overflowing sack of extras for Bill and Marion to take home. Morgan and Shannon busied themselves making a fire on the raised hearth in the little log structure that was roofed with moss-covered cedar shingles on pilings and built out over the bay. Both children loved to be asked to do adult things with minimal supervision, and I knew I could trust them with matches and fire. Their careful tending of the blaze on the flagstone hearth soon had the lid on the pot chattering like a castanet. The white foam of chopped garlic and nettles, bits of red dulse and sugar kelp, white wine, butter, and olive oil swirled in the froth to bathe each morsel in bubbling Lucullan goodness.

April 21 flew by as we hiked a trail I had recently cut to a cliff-top seascape overlook. We admired early cliffside flowers, sampled edible plants, and stopped at a newly found archaeological site. It was finally time for them to go, and we five retraced our steps from the morning, stepping over fresh bear scat that hadn't been on the trail earlier in the day. I helped Bill untie the wings as we talked about the particulars of steep sand and gravel beaches like the one where they had landed.

There was a slight crosswind from the north of about five knots. Marion was busy with a preflight. She would fly back; Bill had flown down. We stood beside the open cockpit door as they both buckled into their seat belts and over-the-shoulder double harnesses. Barely wider than their own hips, the Piper Super Cub with its 150-horsepower engine was nothing if not powerful. In skilled hands, a Cub can fly away with a pilot, passenger, and a considerable amount of gear in compromising conditions like those in front of them. One of Roger Borer's eighty-two-inch props was slung from the nose, offering maximum thrust for the little plane. We stepped back as Marion shouted, "Clear," hit the starter button, and the engine roared to life. Back at the driftwood line and clear of the aircraft's tail I knelt between my little pals and explained to them some of the things going on to help them understand the basic physics of flight. They were accustomed to me speaking to them in quite grown-up terms. I was always surprised later at how much they understood and retained.

Marion firewalled the throttle in her left hand, pushed forward aggressively with her right hand on the control stick to lift the tail, and they hurtled down the beach. We turned our faces for a moment until the blizzard of sand, crab shells, dried seaweed, and discarded feathers blasted by. Then we turned and captured the image of the little plane as it bravely took to the sky. It lingers there still in my mind's eye, bravely banking to the right, turning out over the wide bay. It is a timeless and precious scene. It climbed steeply, and then like a thrown stone, like Icarus himself, fell headlong into the sea with a sigh.

"Oh my God, please, no!" I moaned as I fell to my knees. "There are no atheists in foxholes," I thought, as I asked God to please save them. The plane struck the water with the engine screaming full throttle, and it exploded in a white arc of spray. It disappeared beneath the surface as suddenly as a spent arrow. Only the tail was suspended as we breathlessly watched for it to be swallowed by the sea.

I knelt down again, looking directly into Shannon's eyes, "Listen carefully to what I say, Shannon." I knew I had her riveted attention. She looked at me bravely with complete concentration.

"I want you to run back to the house, pull a barrel seat from the kitchen-window table, and slide it under the radio. Stand on the barrel; be careful not to fall off, hold on to the counter. I want you to take down the microphone from its little clip, hold down the button, and speak slowly and clearly into the mike. You've seen me do this a hundred times. You can do it." Her gaze was steadfast.

"You don't need to be talking to anybody — just say your name, that you are in China Poot Bay, that a plane has crashed in the water, and that your dad is in the water trying to help save the people." I had a glimpse of Shannon as she shot off as smartly as a Roman candle.

By the time I got to water's edge, I could see that Marion had escaped from the cockpit and was clinging to the tail. I expected to see the tail disappear with her at any moment. It later turned out that the nose of the airplane was resting on the bottom.

I ripped off my clothing; I knew from my lifeguard training that I would never be able to swim the distance to the plane and return with Marion unless unburdened by the drag of my clothes. The water gripped me with such a fiery intensity that I thought I might be consumed by it even before I could reach her. I released her death grip on the Super Cub's tail and swam with her to shore against the pull of the tide. When we felt the gravel rise beneath us, we crawled and then staggered out of the ice water.

I dove in again with everything in my being screaming at me to stay on the shore and out of the water. There was nothing more to be done; Bill must be dead after being under water for so long. Like a black-and-white Rockwell Kent block print, Morgan stood like a post on the shore watching.

I reached the tail. Still no Bill. Putting my face in the water I felt the bitter sting of salt and tried to look down the length of the fuselage to the door of the plane. It occurred to me that the time might have arrived when a friend lays down his life to save his friend. I had heard of this happening in military and civilian life but had never considered having to drink from that bitter cup. The water was winter clear.

All of the cold-water drowning manuals stress the critical importance of keeping the face out of the water. It was like looking directly into a blast furnace. I knew that I could never swim down that distance and extricate him from the back seat unless I could be calm, unless I could oxygenate my blood with full deep breaths. At that point I was so traumatized that I could barely breathe at all. Forcing myself to dive down to Bill was the most difficult thing I ever did. I gulped several breaths, tried to rehearse what I knew I had to do, and I left this world for another. I dove down and down, my fingers following the wrinkled fabric of the ruined airplane. I tried to get Bill out of his seat belt and harness. In any other situation, opening the eyes under icy-cold salt water is something one would avoid at all costs. Yet things were clear somehow, perfectly clear in that submerged stillness. I seemed to be operating in a slow-motion realm beyond any time or space. It seemed as brightly lit as a surgery amphitheater. My own death felt perilously close in that otherworldly place.

There was some loose rope suspended, floating idly in the flooded cockpit. Probably these were the very tie-downs that had secured the wings to earth. Now they were holding my friend like an insect in a spider web. Back on the surface with my lungs on fire I was ready to give up hope and take that first stroke back to shore when a thought struck me as clearly as if I had been lying in my warm bed. I must go down again. I had crossed a threshold. Putting my head under that water again was against all reason, but I went down with a clear and driven purpose. I had to try one more time to get Bill to the surface. If I died in the process, well, that was just the way it was gonna be.

I managed to get him untangled and got him to the surface. I stroked for what seemed like forever for shore until I realized his feet were tangled in the lines. Finally I dragged him up the gravel slope. Marion knelt beside Bill, sobbing hysterically.

There was none of that angels and sunbeams stuff, no white unicorns backlit by moonlight. I held Bill in my arms as he died, I cried and I yelled. Again and again, against all hope, I pushed my own breath into his lungs and watched his muscular chest rise and fall in response to my own exhaled breath. The slippery gravel beneath us supported our bodies with a relentless and icy severity.

I spoke to him again and again: "Come on Bill, we can do this, come on Bill, spit it out, spit it out and take a breath, come on Bill, I know you can do this." I was rough with him, I wanted to slap him, hard, to wake him up.

He lay lifeless; I knelt beside him, feeling the life draining out of me into the merciless gravel. We were halfway down the tide line, halfway between living and dying. There was no confused jumble of sun-bleached driftwood, no tall beach rye, no shells of rock oysters and crabs. There were no tracks in the wet gravel that was swallowing up my whole world. There was only that dull ache of wet rocks and my heart's energy draining out of me as I saw Bill's last chance for survival draining out of him and slithering with his vomit down the beach.

There was a little boy watching all of this. He stood silently at the high-tide line as stout as a little driftwood log. This was education that a parent would prefer to spare the child. Morgan, who had himself crossed to the other side and returned, my own son, whose life I had been able to save with my own breath when he drowned in winter, watched helplessly as I struggled with what was by now a corpse.

My good friend Larry Thompson, surely one of the most skilled pilots in Alaska, heard Shannon's brave little call as she stood up on the barrel seat to reach high for the citizens band radio handset. He had just flown in from the dirt strip in Seldovia down the coast, and his plane was hot and ready to go. He checked his fuel and stood by the double doors of the Cessna 206 as he waited at Homer Air for the ambulance to arrive with the EMTs. They hadn't even closed the door when he was making the right turn out of the Alpha taxiway and roared into the air on runway three.

Later, I stood on the beach where all this happened and tried to imagine myself landing his 206 on that steep beach. No way, I thought, no way in hell could it be done; but land safely Larry did. I will never forget looking at the smooth street tires on the airplane and the deep ruts in the sandy gravel and thinking that he was unlikely to get airborne — even light on fuel and with no passengers. He knew somehow that he could do it without killing himself, the two EMTs, and Marion. I knew that Bill was already dead. The big Continental engine roared to life with the long prop blades sucking up enough gravel and sand to permanently disfigure them. Down the steeply slanted sand and gravel spit it hurtled. I could barely stand to watch, for it seemed certain that it could not take off in that small space with such a load and such a soft takeoff surface. It went screaming in a shower of thrown gravel, lurching wildly to the abrupt end of the beach. There the driftwood shore turned sharply southwest between the octopus rocks and Boy Scout Island. It shook and quivered for a moment with no earth beneath its wheels, as if considering whether it would fly or belly flop into the water. It staggered like a drunk from The Ya Sure Club and thank God, it was in the air, bare inches above the water and clawing the air frantically.

I simply have no recollection of what happened next, perhaps because the rest of that day was too painful to remember. All of this of course indelibly marked the children, who witnessed, participated, and healed from the injuries as best they could.

My old Alaska Railroad friend once wrote to me, "Oh the North Country is a hard country and it nurtures a bloody brood." Everyone has seen friends and family depart this earth by one means or another, but the north country, the land of the north wind, the place where dwell the Hyperboreans, is a hard country where death comes in many guises and often it is not pretty. The land and sea draw as freely on the soul in life as they do in death, and there is an exactness to those lines that leaves one both rested and restless.

Loonsong. Photo © Boyd Norton.

CHAPTER 5

Loonsong

I know of a Mount, the gracious Sun perceives,
First when he visits, last too when he leaves
The world; and, vainly favored, it repays
The day-long glory of his steadfast gaze
By no change of its large calm front of snow.

—*Robert Browning*
From "Rudel to the Lady of Tripoli"

Not a single mountain with a calm front of snow but rather a whole flock of silent mountains surrounds the mountain lake camp. You could easily believe that you have landed magically around this emerald beauty after a migration on the north wind, and like a great white bird, add an avian presence to the tranquil scene. The sense of calm they bring to this summit-clad silence is something beheld by few. The mountains, the birds — real and imaginary — and the stillness are undisturbed by humans.

Hidden deep in the Kenai Mountains lies a small lake rimmed by blue-ice glaciers and towering snowcapped peaks. Few of the peaks and glaciers in this wildly scenic place have names. Other than a few trappers from a bygone era and a hiker or two ferried in by a bush pilot, the Valley of Four Glaciers has been little explored. It is known well by only a lucky and determined few. Like a hidden Shangri-La from another land and another time, the spruce-shouldered mountains rise directly out of the profoundly deep lake to summits above four thousand feet. The crystal-clear waters are peppered with rainbow trout that cluster like playful schoolgirls around the mouths of streams. Beside the forests of Sitka spruce, birch, and cottonwood, the shoreline falls into crystal clearness.

The little lake is half-hidden among receding glaciers, fragrant meadows, and braided rivers. Archetypically it represents an idealized sort of Alaska that hides wild places no living person has ever seen. There are uncounted numbers of black and grizzly bears and moose, wolves and wolverines, river otters, and mink. White-blanketed mountain goats gaze serenely down at this idyllic scene from dizzying pinnacles, while everywhere there is the sound of moving water. There are trickles and freshets flashing in the sun and gently murmuring watercourses half-hidden by elbow-deep mosses. There are white-plumed cataracts roaring from high cliffs as they whisper to the valley far below. In salmon season there are bright reds to be found in every clear stream, and silvers nose up their own distinct watercourses, smelling the way to their natal ground. There are even gaily colored arctic char in pools below fern gardens. Hawks, owls, eagles, falcons, trumpeter swans, and even hummingbirds find refuge in this glacier-carved valley that has no roads. My son, Morgan, helped build some of the first miles in this, the first state park created in the years after the territory became a state in 1959. As often as not, he and his crew improved the game trails that had been used by our four-legged neighbors for thousands of years.

Loonsong is my anchor to windward, my storm trysail sheet, my drogue. I watch the changing of the guard as ever-changing cloud shapes advance, billow, and roll over the pinnacles above my little cabin in the wilderness. This is the place of refuge where the silence and the immersion in wild nature restores me. This is the pinspot in the wilds where I can counterbalance the rush and the tumult, the late-breaking news and deadlines with the inner peace that refuges have always offered to the renunciant.

The steep, glaciated mountainsides are the equivalent and thus opposite of the desert oasis, where inside a walled garden was considered by the Tuareg to be a paradise on earth. The wall sheltered the flowering and fruiting plants and the nomad from the scouring winds and the incessant sun, just as my southeast facing cabin windows offer to me protection from the world that is "too much with us." Winds and avalanches, the sound of the cataract through the canyon, the rain on water and wind in birches, the loons and songbirds speak to me with the voice of the indwelling teacher whose guidance is easily lost as I go about the chores of householder, businessman, pilot, and manager.

How clearly I remember the day at the kitchen table in China Poot Bay when a much younger fellow pored over the finely detailed 1:63,000 topographical maps. Before me and writ large was undiluted wilderness as pure and fresh and clean as on the day God made it. I could see the headwaters of the valley whose ocean terminus was visible across the estuary. The scene before me might just as well have told where the pirate's treasure was buried on Longboat Key among the coconut palms.

My eyes wandered over the terrain, mesmerized as if held by a diviner's spell. I considered the four glaciers, the rivers and streams, the avalanche zones, the wetlands, and a little half-hidden lake. Tight contour lines showed steep-faced mountains and still tighter lines showed cliff faces. Following the parallel lines I counted up a mountainside: two thousand, three thousand, four thousand feet. Wow! Up and across the glacier to forty-five hundred feet and higher still to nunatak pinnacles spinning out of the ice. Lofty summits fell directly into the open Pacific to the east or into lakes and braided glacial valleys to the north and south. The Harding Ice Field looming above it all saw sixty feet of snow each year covering millions of acres. Where else on earth, I wondered, could such magnificence be found with so many hundreds of square miles without a single trace of human presence?

The map might have lacked the full spectrum of color that the land itself contained, but the blues, greens, browns, whites, and blacks gave me all the colors I needed to imagine myself right in the middle of that magical-looking valley. I could feel the gentle rocking of a canoe on that lake. I could almost see the bears and moose, sheep and goats, wolves and wolverines. Enchantment, pure enchantment, streamed up like a radiating light from the face of the map. Accustomed to reading the smallest cartographic details through training as a young Boy Scout, my eyes must have sparkled when I saw that the large glacier at the head of the canyon did not drain into the small lake but rather skirted it on the northeast shore. I knew the significance of this fact but had no clue how critically this would play out during the rest of my life.

In the late 1960s, I was fully engaged in acquiring all the skills that I thought I would need to prepare me for the life of a lodge owner, bush pilot, big-game hunting guide, and commercial fisherman. Diane was raising a first child and about to have another, tending a garden, putting up meat and vegetables for winter, and keeping our one-room log cabin functioning like a Swiss watch. There was no such thing as ecotourism, I didn't know anyone who owned a kayak, and we had never heard of anyone living the kind of lifestyle we idealized. The hearty folks we met who approached that pattern were a breed apart; we admired them. I set my sights on a master guide license, but it would be many years before license no. 51 had my name on it. Even at that time we could see the disconnect between ecotourism and big-game guiding.

A year after poring over those maps, I flew as a guide with a hunter up the valley that was headed by four glaciers and had my first sight of the little lake. The emerald-green jewel hugged tight against the side of a kelly-green mountain. It was a storybook setting in a wilderness crown not far from the blue-ice terminus of the glacier. Even from above, I could see that its waters were gin clear. Serendipity would soon enough acquaint me with the deserted trapper's cabins beside the lake.

We planned to hunt mountain goats high in the foggy peaks, which today can be seen from the cabin we would eventually build beside the lake. I had spent time in the Alaska Range south of Denali at the headwaters of the Kuskokwim River with Warren, a former Green Bay Packer, now a contractor and my client. We had taken an enormous grizzly over a moose kill and established a brotherly friendship of mutual respect, so before parting company, he gave me a deposit for a goat hunt confirming his intention to return the following year. He wanted to join Diane and me in China Poot Bay to experience all that I had told him about: crabs in crab pots, shrimp in shrimp pots, a rookery teeming with seabirds, seals, sea otters, and sea lions as well as fishing, hiking, and sea kayaking and the view from our living room of the smoking volcano.

We landed on an alpine lake about two thousand feet above the valley floor and the lovely lake I had seen. We unloaded at the leading edge of one of those nasty fall storms for which these coastal mountains are famous. We had good equipment and knew how to use it to best advantage, but as the storm intensified, we were barely able to stand up to the katabatic winds. Visibility was limited to a few score feet in freezing rain and snow. One wishes for a cave or the lee face of a cliff for protection, but there was no hope of shelter. We huddled in the tent so that our combined weight would keep it from blowing right off the mountain, and I brewed cup after cup of tea and soup with a tiny mountaineering stove. We had caught a glimpse of a wolverine loping along the side of the lake when we arrived. It offered a segue into stories that would while away the hours and divert the hunter's attention from our predicament.

The sturdy mountaineering tent served us well, but day by day we got wetter and wetter as the relentless storm intensified. I had run out of wolverine stories. One of the titanium tent poles shattered and then another. We were in trouble. I told Warren that if we could get down to the timber I could get a fire started. More than once in the past I had been able to collect enough spruce sap to set dripping-wet firewood to flame. The rugged Green Bay lineman finally had had enough, and we staggered off the mountain, slipping and sliding on the ice and snow across precipitous slopes, in constant danger of falling to our deaths. We reached the lake as the storm raged on and found a dilapidated little cabin tucked into the forest beside the lake. It was a shambles with a hole in the roof, the floor partially rotted, and the stove and chimney in bad shape. Bush repairs were made to the stove so as not to burn down someone else's property. Soon there was heat and salvation. A pair of loons ventured close to ask with their mournful cries what sort of creatures had come to their lake? Wet, stinking, and miserable was our answer to them, although we had found a refuge that couldn't have been matched by a room at the Hilton. There were no goat steaks to fry, but soon there was a hot soup sliding down our gullets and warming us from the inside out. We were going to be okay after all.

In the early years of statehood, Alaska offered opportunities for those seeking to acquire a little piece of the wilderness. The state's Homestead Act offered up to a quarter section of 160 acres with flexible provisos for improvement and habitation. Smaller sites of up to five acres could be acquired through the Trade and Manufacturing Site process. In the 1940s, three pioneering homesteaders from Homer took advantage of this opportunity and did a hand survey of five acres each at the east end of this lake. Those lucky trappers would have had this great game-rich valley to themselves. They built the minimalist board-and-tar-paper shelters, not unlike the one-room log cabins replicated all over Alaska as trapline cabins.

They brought in the basic materials in a small airplane on floats in the summer or on skis in the winter. The sizes and shapes had to conform to what could be tied to the struts or spreader bars or put in through the small door and smaller space inside.

Wisely they flew in a surveyor to formalize their recording, and that staking would in later years became a benchmark for surveys throughout this vast and remote area. To this day, the permanent brass caps stick proudly out of the forest floor saying that these three adventurers gained title to the land during territorial days.

When Warren and I had finally made it back to the lodge, Diane listened to the story attentively as she busied herself at the woodstove preparing a meal for us. I considered her smooth efficiency as she bent to pick a piece of wood, flipped open the firebox, popped it in, and as quick as a wink was back to her work. She paused to look up and consider my effusive descriptions. I continued, "Well, my dear, the mountain rises directly out of the depths of the lake; the loons have probably nested for untold generations on the little island in front of the trapper's cabin. They came to check us out and actually seemed to be looking into the windows. They swam really close and then began to softly call. Then they ran wildly across the surface of the water shrieking like crazy people. I wish you could have heard their cries echoing off the surrounding mountains. There were several goats on the cliffs where the lake drains into the glacial river. There was a freedom and freshness to the place that was nothing less than magical. I felt as if I had found something I had been looking for my entire life." There in Diane's kitchen, with those descriptions, we began to call the mountain lake camp "Loonsong."

I had been smitten by that lake just as I had fallen head over heels in love with China Poot Bay years before. It seemed clear that this would be the perfect place for a subalpine extension of the lodge. In one visit, our guests would be able to experience both a coastal estuary and a mountain lake. It was a silly idea for many reasons. We had no money and more than enough challenges in front of us as we tried to build our home and lodge on the coast. Even if we could have purchased the property, what sense was there in owning land for which we had no means of access? Then and now the cost of flying to Seattle is about the same as the cost of the bush-plane flight to the little lake. What hope would there ever be for getting materials to the lake when there could never be a road? The idea of hauling supplies in winter by snow machine and sled was likewise impossible given the extremely rough terrain and the many river crossings required between the ocean and the isolated lake. One side of the lake was regularly scoured by thundering avalanches, and the other was guarded by a narrow slot canyon where the glacial river blasted through the near-vertical walls. We had owned a fine young gelding and would soon own a winter ranging stallion, so I considered the broad grassy meadows I had seen beside the lake. Perhaps we could develop the place using packhorses? I knew just enough about horses to know that this was folly.

Back in town, I tried to find someone who knew about the lake and who owned land there. One of the area's first pioneers, Dick Inglima from the grocery store, seemed to know everyone and everything about the Homer area. He directed me to Sam Gasparak, owner of the Club Bar, which was for many years one of Homer's most popular gathering places. At a time when everyone in town knew everyone else, it was a place to have a birthday party, a setting for music and dancing. Sammy presided over the place with small-town good cheer. He was liked by everybody.

After years of regularly pestering him, however, he did agree to sell his five acres for $1,000 down and $100 per month at 8 percent, for a total of $10,000. It was a huge sum in those tight times, and the purchase made no practical sense at all. Diane and I had no regular paychecks, and we had strapped ourselves into mortgage payments on the land where our hopes for creating a successful lodge were still tenuous. Added to this, she was pregnant with the second baby, and remote land acquisition didn't make good practical sense. God bless my little wife, she agreed to this craziest of schemes and the dream of building a still more remote extension of the lodge. More than once in any kind of exploration in science, technology, or frontier pioneering, tell someone that a thing is impossible and it will get resolved. Get a good woman like Diane behind it, and it will not only succeed in short order but also prosper.

Without an airplane or the money to charter one, there would be many trips to the bare-bones camp on foot or skis in the years ahead. These were restricted to winter when the several river crossings were made possible by the ice. On one of my many long midwinter hikes alone, I fell through the ice where the river ran through that vertical walled canyon located about an hour short of the cabin. The boreal regions claim scores of hikers every year in exactly this way. It was well below zero in that winding canyon that hadn't seen direct sun for three months. In a matter of minutes, I was imitating the Tin Man of Oz in my insulated Carhart cotton coveralls. My arms would barely swing at the elbows or shoulders and my knees would barely bend, so solidly frozen was my carapace. My situation wasn't funny, however, that was for sure. The last few miles to the cabin required climbing an almost vertical wall of rock that I traversed by pulling myself up on spruce branches. It was a challenging scramble even in summer; now it was as slippery as a wet bar of soap. On the verge of pitch-dark, I staggered exhausted through the door in the first stages of frostbite. Shaking badly, I could barely hold a match to light a fire and burned my fingers in the process. Thank goodness I had left the obligatory dry kindling splits, paper, and cardboard at the ready in the off chance that someone would come upon the cabin in an emergency. As far as we know, in more than forty years of occupancy since those first visits, not a single person — except family and friends — has ever walked into our remote property. The fire was soon drafting handsomely, and as I began to thaw, I created a large puddle on the floor that refroze right away. That night I could sleep only a few hours at a time. It was essential to keep the fire going to dry those coveralls before the dawn hike back down the valley to check my traps and finally get back home to the lodge after dark.

By some curious twist of fate, which included a serendipitous meeting with a space scientist at the University of Alaska in Anchorage, we showed up on the computer screens of the National Aeronautics and Space Administration. One of their astrophysicists from Los Alamos, New Mexico, had taken a sabbatical from his regular job at the space center and was in Alaska to document the pioneering spirit that was still alive in America. For his book on the subject he wanted to capture the character and personality profiles of people who would be chosen for long-term space voyages. He observed that there were still a few people alive who had gone west in covered wagons and had in other ways pushed the frontiers of geographical discovery. He was interested in those who pushed the limits of other kinds of frontiers as well: math, chemistry, and human biology. But he was most interested in those who reached the physical limits of a place and wanted to keep going. Eric Jones, author of *Colonization of the Galaxy*, came to China Poot Bay with his friend Mead Treadwell and with his tape recorder and notebook. We hung spellbound on his descriptions of why NASA thought that sooner than later, people would indeed leave our solar system as colonists on one-way trips into the vastness of space. I tingled with excitement at his descriptions. He called to mind how I felt watching my college physics professor at the blackboard chalking out the mathematical formulas describing how a vehicle could reach escape velocity to leave the solar system and then achieve something approaching the speed of light.

"Okay, Eric," I said, "I'm ready to sign up. When can I get fitted for my space suit?" I was — and am — ready to go, much to Diane's consternation.

I went on. "Eric, there are two things that might compare to the isolation and loneliness of space. One is diving beyond one hundred feet with scuba equipment in places where no one else has ever ventured. The other is being in a remote trapper's cabin in the Alaska mountain wilderness in the dead of winter." He took careful note, and a few weeks later he chartered a helicopter, landed at the lodge, I climbed in, and up the Valley of Four Glaciers we went. The chopper hovered over the ice; we jumped out and were left alone in that great white silence. We noticed large tracks in the deep snow near the cabin as we split wood and witnessed the ghostly fading of the light as dim shadows moved in the forest. That night the wind howled and rattled the woodstove chimney as the fire burned itself out. In our sleep, we heard wolves calling, their ghostly howling ricocheting between the snowy peaks and distant galaxies. Occasionally there was the deep basso rumble of the lake groaning in the star-filled void. The only earthly similarity to that sound is the slow forward movement of a hundred-car freight train as the massive steel coupling mechanisms grind and moan. We felt very alone and vulnerable. The thin umbilicus leading back to the familiar world was a helicopter whose receding turbine engine had left us in absolute white silence.

In the morning, the wash water was frozen solid in the chipped enamelware bowl. "This," I said to Eric, "is a tiny taste of the loneliness and fear that must be a part of leaving everything familiar behind."

The solitude of a remote mountain place fills one with companionship and another with dreariness. For me, there is no room for solitary thoughts when you are surrounded by brother mountain and sister glacier and the animals are your many children. The night sky is your lover and the stars the beloved. There is no room for boredom when the splitting maul and woodpile ask for your attention, and there is that list of chores taped to the inside of the cabinet. "I love not man the less, but Nature more" was how Lord Byron expressed this lure of the woods. In such a place every change in the breeze and every falling leaf is the beginning of a new direction. Every birdcall, every spreading dimple of a rising trout, every glance up from my desk across the cove is a celebration and a gift.

Diane's kind father, who had been supportive of our dream, died not long after we purchased the land at the lake, leaving us a small acreage in Illinois where her parents had hoped to build a retirement home. Diane's mom gave us the land; we sold it, and spent half of the money at the lumberyard and the rest with Bill and Barbara DeCreeft at Kachemak Air Service. With the right weather conditions and light on fuel, we loaded their beautiful 1958 600-horsepower de Havilland Otter time after time, over the course of the next six years, with maximum loads of twenty-two hundred pounds of materials to build our Loonsong Mountain Lake Chalet.

Well more than one hundred trips were flown to carry more than a quarter million pounds of freight to support the construction of our mountain dream. Two-by-fours and plywood, cement and rebar, metal roofing and tar paper, couches and stoves, cabinets and propane refrigerator — all the trappings of a remote but comfortable mountain chalet. The job took more than six years and an untold amount of work by several dedicated and capable workers, some of whom toughed it out year-round.

By now the lodge business was humming along pretty well and the need for our own airplane became more and more obvious. Although years had passed since I flew for hunting and fishing lodges in the Alaska Range and north of Dillingham and McGrath, I kept my oar in the water and my skills fine-tuned by flying rental aircraft whenever I could. For years I had been adding to a little savings account at the National Bank of Alaska named the "Steinway account." I had had a love affair with those grand pianos since my childhood and considered ownership as a pinnacle of success. Every time there was loose change or a windfall of cash, it went into the coffee can with the piggy-bank slice in the plastic lid. I actually nurtured the fantasy of taking a Steinway to the mountain camp slung under a helicopter. I liked what Robert Browning said about such fancies: "Ah, but a man's reach should exceed his grasp, or what's a heaven for?"

The years passed, the savings account grew, and my foolishness was finally resolved when a Piper PA-12 with a 150-horsepower engine, Super Cub landing gear, and Atlee Dodge skis came up for sale at a price I thought we could afford. Back in the cockpit, I was romancing the peaks and ridges above Loonsong and out over my beautiful Kachemak Bay. Without floats, however, the plane was of only marginal value to the business. When a pair of 1951 EDO 2000 floats came up for sale at a price that matched the Steinway account, I abandoned my piano dream. Its FAA registration numbers were N3580M, so we called her "eight-oh mike," "buckshot mike," and even "eight-aum-mike," with a family pun directed at my yoga-teaching certificate. Since Lindbergh named his plane *The Spirit of St. Louis*, I thought my lovely aircraft should have a name too. I attached to the fuselage just behind the engine cowling a handsome decal exactly as seen above the keyboards: the signature treble clef and the word *Steinway*. It was just grand.

Loonsong Lake as seen from the window of Steinway *eight-oh mike. Photo © and courtesy of Michael McBride.*

I made scores of winter deliveries of supplies and groceries on the ice with eight-oh mike on skis, sometime landing, and other times, when the ice thickness was not reliable, air-dropping supplies. There were countless other trips on floats hauling lumber and canoes and kayaks tied to the float struts. The sturdy workhorse was happy to carry pairs of propane bottles weighing 186 pounds each, one strapped to each float. The PA-12 was one of a small handful of Alaskan bush planes that were highly sought after by people like us working in remote places.

After years of challenges overcome one at a time, the comfortable accommodation on the lake built beam-and-post style with huge spruce logs was finally ready for the first guests in 1989. We worked hard at drawing the attention of writers and photographers, whose promotional efforts we hoped would attract clients.

When not working at the lodge, Diane and I flew personally with clients to the mountain camp, where every year loons nested on the little island in front of the cabin. Several times per summer, a floatplane would pick up guests in front of the lodge, give them a world-class flight over the glaciated peaks, where they usually saw bears, moose, and mountain goats, and deliver them to find Diane and me on the dock under the spreading cottonwood tree with our Piper PA-12 tied to the shore. There would be smoke coming from the sauna, fresh flowers on the table, and wine chilling in the spring box. After a dinner of fresh salmon and herbs from Diane's garden, we paddled the refurbished 1929 Old Town canoe that had been given to us across the lake to the now refurbished trapper's cabin, thus giving the one or two couples or family the privacy of the grand facility. In the morning, we paddled back, lit the fire in the living room woodstove, put on the coffee, made breakfast, and offered an exciting range of guided hikes, paddles, and fishing adventures for the day. Jose DeCreeft, son of the pilot who flew in the materials for the grand building, often returned to take our guests brown bear viewing for the day or for hikes radiating out from lakes in the glacier country far above tree line.

Loonsong was becoming more than a take-your-breath-away-scenery kind of place, more than a cozy refuge from the rest of the world; it became the axis of our hearts. For me it became the very center of my being. Even when I am not there, it comes to me in dreams. It shows itself in a myriad of forms, but always there is the same theme: pushing away from shore to venture out on the lake in the star-sparkled darkness in the Old Town canoe to partake in a festival of the gods.

In the fall, there are many reasons to make the short crossing of the lake, but my favorite provocations are the cottonwood-edged meadows behind the old trapper's cabin. There, the finest patches of highbush cranberries are found just before the lake freezes over.

Here in the Far North, where the climate does not support native fruit-bearing trees, the ground sends forth in their stead a bounty of delicious berries. We can count thirteen varieties around China Poot Bay and at Loonsong.

A poet I once read, whose verse I can no longer quote, spoke of how smells can "make your heart-strings crack" with the tremendous power of recollection they can trigger. None of our other four senses holds the strength and associated emotional response provoked by smell. When opening one of our highbush cranberry jars, there comes an overflowing redolence that recalls another dimension and the very essence of the wildness. With a slow inhale comes the woodsmoke rising from the sauna, the damp smells of the mosses at the edge of the lake, and the fullness of icy-cold water tumbling down the mountain and into the sparkling meadows. The berry smells bring images of frosty mornings, vapor rising from the lake, sleeping under heavy blankets, the sharp crackle of the kindling splits as the first fire of the day in the woodstove bursts into flame.

In this peaceful place, loons, bears, squirrels, moose, wolves — even the fish under the floating dock — know that I am here. The smoke from my woodstove wanders up and down the valley as unassuming as a ghost but communicating as clearly as the daily paper that a person is here in the valley. Those neighbors who have shared the same trails with me know a good deal more about me than you would expect, because any time I stop to urinate or sit and rest, it becomes a focal point for their attention as they pass. Urine and fecal deposits are to the animals what business cards are to the coat-and-tie people — they tell you the basic facts: male, female, profession, workplace, and what you ate at your last meal.

Wild nature plays with our intellect and our sense of wonderment at times like these, and we spin down like DNA spirals into our own subconscious and simultaneously burst like a Roman candle with bits of our being hurled out among the stars. Standing completely grounded with feet on the earth in wild nature, we can find balance between the two vast extremes and there experience a serene calmness that is commonly thought to be the provenance of only saints and avatars.

The rugged smell of moose meat panfried in the cast iron skillet with nettles, olive oil, and garlic filled my soul and overflowed into every nook and cranny of the cabin. The fire in the woodstove crackled and popped, its warm glow radiating throughout the room. The dishes were washed and stacked to dry. It was time to paddle the canoe out into the darkness under the stars, and I was as keen as mustard for the feeling that I knew would greet me there. I prepared to push the canoe from shore into the darkness where the two of us would ghost silently into the dark. There was a slight crunching of gravel as my right foot pushed against the shore and launched the canoe into the lake.

The transition from shore to water is made without hesitation; there is no quarter for timidity. There is a brief, almost imperceptible moment when the body is neither on the shore nor quite yet in the canoe. My single push from shore takes us on a long glide that seems to go on forever. Stars peer curiously from above the mountain's avalanche-prone face. In the water, I continue with a mindful J-stroke, and the canoe responds as gently as a lover to a caress. The water below the keel and close to shore is profoundly deep.

This bright creature of flashing wave and dappled sunlight is subdued now, tranquil in the meager starlight. She trusts me as Chenik, my black Lab, trusts me to keep her safe, to show her the way. She is my beloved, she cradles my vagrant dreams. She comes to me at night, awakens me gently, and spirits me away to magical and enchanting places. Tonight we might traverse a galaxy, travel at the speed of light, become gods ourselves as we glide about among the stars bright overhead and whose reflections are spattered all around us. They bid us "come forth." We can and we do.

The canoe and I are fused so seamlessly together that we know one another's moods as surely as the dog knows the master's mood even before the ritualized evening walk begins. I know of no balm, no gentle bath, no caressing touch, no blessing more calming and restorative than that freely given by this curvaceous waif who waits patiently for me inverted and suspended over the living room, where I have designed a simple system for raising and lowering her without assistance. The honey gold of her ribs spreads a warm glow over the room. She is always there, looking so lovely, poised at a moment's notice to accommodate herself to me. Our launch into star-speckled darkness is but a simple act, although it is has never been experienced by another, ever, on this lake deep in the Alaskan wilderness.

When you push away from shore in a craft of any size, you do so with obvious purpose. Why else would you leave the relative safety of terra firma to bet against the odds that you might again return safely to land? This seemingly simple act has throughout human history been the elemental threshold. After that, it was flight and then the ability to stay under water for longer than a breath hold. River, stream, lake, pond, or ocean, crossings in birch-bark canoe or ocean liner are adventures of the first order. To anyone else, my craft and my mission were as inconsequential as a single raindrop in the vastness of the ocean.

Poets and troubadours across the ages have written and sung about the touch of sadness that comes with late fall and the approach of the first snow. This is something very different from the dancing and singing that comes when the fishnets have been stored or the potatoes dug in the midst of a bountiful harvest. This is the change being forced on us by the sun, who every day sets a little farther to the south, leaving those of us in the Far North to deal as best we can with the abbreviated visits of our brave friend Helios.

I flew on floats to the hidden valley in November again that year to spend the last precious days before freeze up. My goal was writing, but it was the splitting wood and hauling water that were to be the backbone of the retreat. I make the pilgrimage yearly if I can, drawn like a magnet by my senses. They overpower the brain, so strong is the desire to be in the embrace of wildness. When I arrive, there is a baptismal ceremony. I lower my face to the surface of a full bucket of water freshly dipped from the lake and, communion-like, drink my fill. This simple act is as deeply satisfying as feeling the shore under your feet after a long swim far from the coast. The water is so cold it will make your teeth hurt. There simply is no other taste like that. There is no smell like the spruce smoke scenting the early morning air as it curls from the sauna chimney. Most of all, however, I just want to savor that mysterious feeling that comes with watching the first snowflakes fall on the mountain lake.

There is a certain stillness that descends on the narrow valley when snowfall is imminent. It creeps imperceptibly into the shadow of the mountain on the furry feet of a lynx. If it comes during the night, it ghosts through the cabin walls such that even during sleep, the profundity of the mountain silence gently awakens you. Without seeing, you know that the flakes are drifting steadily down just beyond the window, so you contentedly burrow deeper into the warm folds of the woolen Hudson Bay blankets. Gradually Morpheus will release her gentle hold, and while it is still pitch-black, you will rise to a world newly baptized. As you step out of the cabin and into the fresh snow, it is like stepping out of a bath, clean and refreshed. There is a distinct smell that comes with that first snow too, and it splashes over all of you like an arching wave.

The sense of impending snow may also come in daylight while hiking a trail, hauling water, or splitting wood. The moose hear the sharp crack of the splitting ax on the chopping block and reply by rattling their horns against a stout branch. There is actually a texture in the air announcing that the time of falling snow has arrived. It is as if an invisible covering draws around you and offers to muffle even the sound of your own breathing. It brings a certain mysterious satisfaction that one cycle of life is ending with the season and another is about to begin.

It is the great depth of our lake that prevents it from freezing until late in the winter. The shallow lake in town by contrast usually freezes up by late October. Then the airplane's floats are traded for wheels or skis, and all the lakes and rivers become runways where landings might or might not be possible or safe. Once the winter cold has set in, the rivers in our remote valley shrivel up and freeze to the point where it becomes possible to hike or ski across them, but we are the only ones who ever do. There is a period of a month or more when the lake in town is frozen, but the mountain lake still has open water. This is the time to follow the fresh tracks elsewhere and read them to learn about the affairs of wolf or lynx, wolverine or weasel. In a good year, there is snow enough to cross-country ski those long winter miles to the cabin.

It is unheard of to be in the mountain valley this late in the year with a plane on floats tied to the dock. The curtain is poised to fall on the lake in town, freezing it over in a single night and closing it to floatplanes for another season. That year, the unseasonably warm temperatures continue to be fickle, and the mercury errantly bobs up and down. So far, the snow has fallen and retreated several times, but I have not been here to witness it. The lateness of the season makes it abundantly clear that winter is past due, and that I'm sticking my neck out being here. If the lake in town freezes while I am here, I will have to land in the open bay and taxi into the boat harbor. At this time of the year, the prevailing wind is from the north, and such a wind often makes the water at the harbor's mouth too rough for landing. When at last the bay is calm and I can land outside the harbor, I'll still have to face the complexities of getting the plane out of the water and trailered up the highway into town, where finally the mechanic and I can remove the floats for the transition to wheels or skis.

For the past month or more, every fiber of my being was focused on flying to the mountain lake camp. I've been waiting and watching as the snow line moves down the mountainside day by day. Business and family affairs seemed to be conspiring against getting here. There are always more important things to do, it seems, than retreating into the wilderness. I'm here against the weather odds too, so I feel like the little Hobbit, sneaking the jewels out from under the sleeping dragon. As the seasons change in the Alaskan glaciated mountain country, the weather is nothing if not fiercely unpredictable. A mountain lake facing a glacier in a narrow valley is no place for a floatplane in November unless the pilot is willing to take chances with the safety of the tethered craft. But I'm willing to take that chance to be here and witness again that ethereal first falling of snow on the lake.

A look out today's window gives a small insight into the larger forces at work. The great weather blender is turned up to moderate speed. The radio recites a litany of gale warnings in Shelikof Strait, high seas in the Barren Islands, and "fifty knots out of bays and passes; Kodiak Island reports horizontal rain and fog." In this kind of pattern, the local glaciers become only bit players in a much grander drama. At this time of year, when the seasons are in transition, huge weather patterns come swirling out of Siberia, across the Bering Sea, or roaring up the Aleutians and have their way with Alaska.

Finally, I make it to Loonsong. Rising in the predawn darkness to an ice-cold room deep in the mountains offers a sense of contentment without equal. In the little cabin's pitch-blackness, a single star beyond the frosty window winks me awake while I fumble for a match and light a candle. There is no need for the jarring rattle of an alarm clock, for I have once again lapsed into the rhythm of this north woods valley. Sleep in this good place is as Shakespeare described it: "sore labour's gentle bath, balm of hurt minds." Here the moon puts me to sleep and the stars gently prod me awake.

Just for a stolen moment, though, as the fog of sleep clears from my brain, I draw the covers up to my chin. The little child inside smiles mischievously and listens attentively. It's the sound of hundreds of waterfalls hurrying about their business, down the mountainside on their way to the lake. There is no wind, and the murmuring water is the only sound. The cozy hole under the down comforter is sharp contrast to the heatless space of the little room. Getting up is no languorous yawning or sitting on the edge of the bed, slowly rubbing sleep from the eyes; this is a leap over an abyss, a plunge, a flurry of activity, a rush to cover bare flesh. This is the true calling of recycled plastic containers that have been converted to high-tech fleece. This is the moment when polymer technology becomes the best friend of chicken-skin nakedness.

The star that prodded me awake is firmly mirrored in the still lake just beyond the window. The candle flame is mirrored back at me, mimicking the star. I peer out, nose smudging the glass. I yearn to see the aurora this morning, dancing above the mountaintops and glacier. There is a faint pinkish glow above Wolverine Ridge to the north. I look south for comparison to a darker portion of the sky, then back to the north. There is just the slightest hint of color in the sea of ink; yes — there is the faint glow below Ursa Major. This would be a fine morning for northern lights, that most sublime of heavenly displays. I have seen it scores of times in a vast assortment of conditions; for days the radio has been bringing reports from the astrophysical lab at the University of Alaska that there may be once-in-a-century viewing opportunities on the way. First things first: there will be the sacred ritual coffee.

The flickering candle, the fleece, and I climb the stairs from the bedroom on a lower level in slow motion to the top of the landing. I'm barefoot and pause for a moment outside the sliding door. Beyond it the temperature changes from the very cold of the outdoors to coziness. There is a distinctive rattle as the pocket door scuttles into the wall's recess. This is the defining signature of my morning; there is no other sound quite like it. I step from one world into another like a diva striding onto the stage in grand entry. The welcoming space springs to life with the single candle, and I feel like Howard Carter entering Tutankhamen's tomb surrounded by golden treasure. Here is presented a solid room of huge logs, a many-cushioned sofa, a Kennedy rocker, and picture windows revealing stars over glacier, mountains, and lake. There are rugs from Afghanistan and Iran next to rugs woven by Diane's mother. This is a home in which to live and love and grow; a cradle for dreams and life. This is not a cabin of inactivity used only during trapping season or as a place to escape from wife, family, or business. This room is alive with spotting scope, musical instruments, a heavy circular dining table with legs carved like a lion's paws. Friends and family can eat and drink and talk long after meals at a table like that. There is art and sculpture and a grand library of books, including the Five-Foot Shelf of Books — the Harvard Classics — and a complete set of *Encyclopaedia Britannica* and scientific texts to boot. Colors and textures and smells playfully intertwine. Living room, dining room, and kitchen are handsomely arranged in a twenty-four-by-thirty-six-foot plank-floored refuge. It is Spartan and simple in the way of remote places, but there is a sumptuousness produced by many years of loving occupation. While there appear to be all the accoutrements of a "modern" house, everything came in by floatplane, and this presents its own distinct feeling. If it can't be packed in on your back or squeezed in through a floatplane door, it's not in this room. I can hear my lifelong friend Bill DeCreeft saying, "If you can fit it through that door, I'll fly it anywhere for you."

The woodstove lives in the center of the room, where it sits like royalty on a slightly raised hearth. Its generous throne is handsomely tiled lest burning embers scar the floor. Honey-colored Douglas fir planks surround the broad hearth and accent the room. The joy of singing woods is everywhere celebrated.

Apparently the "sleeper log," which I dutifully placed on the coals last night, burned slowly. The damper on the stove was turned down tightly last night and then opened a tiny bit to give it just enough air to burn slowly, silently all night. Now turning it fully open, I laugh as I say aloud to the stove, "Let her rip." Probing with the piece of steel rebar in the ashes and charcoal, I disturb unburned wood embers; they glow like a dragon's eye. Leaning forward so that my head is nearly inside the stove, with a few focused breaths I rekindle the flame. Fresh splits begin their musical crackle, and the day is now formally open for business. Audibly sucking fresh air, the woodstove surges to life. The door is left ajar while coffee gets organized. The auroral glow has increased beyond the picture windows.

The ancient ones, be they Sugpiaq or De'naina, surely knew of the abundance of mountain goat and black bear around the lake that we would come to call Loonsong. We regularly see them on the mountainside that rises directly from the edge of the lake to more than four thousand feet. The cabin lies just across the creek and its little bridge from the sauna, and we have actually had mountain goats look curiously into the windows. Each spring when I am opening the building for the summer, I climb up to the sauna roof to check the growth of the willows and cottonwoods I planted there among the seed-thrown spruces. I take a quick scan, looking for goat droppings. I cherish an image of a mountain goat on the roof of the sauna contentedly munching the tender shoots in early spring when the high slopes are buried deep in snow and when avalanche thunder is a daily occurrence. In winter when the leaves are gone, the observant eye might catch a glimpse of a wolf or wolverine as it steals silently through the dappled camouflage and slanting light at lakeside. Doubtless the old ones hunted these crafty residents on four feet and added their furs to their colorful and utilitarian clothes. Wolverines held then, and hold now, a mythic reputation that provided the hunter and trapper many fine stories with which to regale his fireside listeners. There remains a palpable presence of these long-gone people, something in the air perhaps — something in the slanting sun through the rising spirals of smoke from the sauna.

When digging the pilings during construction of the sauna, I found charred gravel at the base of the hole. Was it evidence of an ancient campsite fire? We would like to believe that this place gave the people the same quiet joy that it gives so freely to us today. In that ritual bathing-surround of the sauna, with the sod roof above and the charred gravel below, one is spirited away to another time, another dimension. Our thrown water hisses against the hot rocks, and steam rises like a sacred communion shared across the millennia.

Native elder Dora Mulch harvesting a bounty of berries. Photo © and courtesy of Michael McBride.

CHAPTER 6

Apanuugpak

The Native Connection

> *"I came down by water,*
> *I will leave by air."*
>
> —*anonymous*

A wild Arctic storm was raging just beyond the door of Beda Slwooka's little house. Her ancestors, the Kingugaaghhaq family and their predecessors, had lived for thousands of years on the shores of the Bering Strait on what are today the American and Siberian coasts. She was telling me an ancient story as we sat on her floor in the village of Sivuqaq, which means "snow shovel"; it also means "walrus scapula." This is the name of the island, the mountain behind the village, and the village itself. White people call it Saint Lawrence Island and call the village Gambell.

The storm was pushing the early winter ice out of the Arctic Ocean and down the strait. In the space of a day and driven by those ferocious winds, hundreds of miles of open ocean would be covered

— frozen over for the rest of the winter. On that westernmost tip of the island twenty-six miles from Siberia, unimaginable pressures came to bear as the open ocean ice was compressed between the opposite shores.

Millions of seabirds and ducks, whales, and walruses fleeing from the advancing ice edge would be displaced. The urgency of this utmost northern transformation literally shook the house. We were skinning the white arctic foxes that my hunting partner, Yemegaq, and I had taken the day before. Sitting straight legged, weaving nets or grass baskets, skinning and butchering game — it's all done sitting on the floor in the central room. *Qasqi* is an ancient name for such places where, while working, instructing, dancing, or singing, one tells stories to pass on the ancient ways.

"In the culture of my ancestors," she began, "the greatest gifts we can give or receive are those without material form: songs and dances and stories." And with that she began to tell of her ancestors.

In those moments Beda Slwooka and I were free from the cares of the wider world, immersed as we were in the tasks at hand. The left brain relaxed as the tactile senses came alive. The earthy smell of blood and sinew was clean and light on our hands; our minds were fully open to another dimension as the ancient stories unfolded. I felt an all-encompassing peacefulness that some seek in a place of worship. Her voice was calm and soothing. The ragged edge of a November storm pulsed through a window. Basic, elemental, even primeval was this most ancient of settings.

When Diane and I were new to Alaska, we could see that we had a lot to learn from the old-timers who had come into the country in territorial days, but we could sense that it would be the Native people from whom we would learn the timeless things that we sought. We recognized that we were witnessing something very special, a historic transition that would occur only once. The indigenous people were in the final phases of transitioning from hunter-gatherers of old to modern corporate citizens. While the white people were commonly called pioneers — although strangers in a strange land fit our concept of storybook pioneers from any era — it was the Native people we encountered who were something else again; we saw them as living treasures. They had inhabited this land long before the birth of Christ or Confucius, long before the first pyramids. Those survivors of the Bering Land Bridge had hunted for the Russian *promyshlenniki* who set up trading posts in their territories, as the first Russians moved up the Aleutian Islands to the Alaskan Peninsula and into the mouth of the area where we chose to make our home, Kachemak Bay.

The real pioneers in Alaska in the fullest sense of the word were of course Yemegaq's people and those first coastal migrants to populate our own shores in the Kachemak Bay watershed. I like to daydream about an ancient time when the land was completely without people. It is interesting to imagine for a moment who exactly the first people might have been, what their names were, and where they were living before they ventured beyond the limits of what they knew, to this new place. Imagine the weather on that day, their clothes, and the expectant or fearful looks on their faces as they surveyed the unknown shores ahead. There was no one to tell them, those eastward-moving people, what to expect as they paddled around the next point or gazed at the islands in the distance. They might have carried a hawk or owl feather before them to propitiate the unknown spirits of this empty place. Over those several thousand years, different peoples with different languages, cultures, and patterns of tool use came and went along our section of the Alaska coast. There was even a thousand-year period when our region of Kachemak Bay was abandoned entirely.

The descendants of the First People to live in Alaska are still living in our midst, and we are living on their land. Only by standing on the solid foundation of this deep understanding can we build sustainably for the future with the humility and boldness that is required in this challenging place.

Who am I by contrast but a recent arrival that drove the Alcan Highway in an old pickup truck, enclosed in glass and steel propelled by fossil fuels. I pushed a button to get warm air on my windshield and stopped to get gas and a Coke. I put quarters in a pay-phone slot to call home. I like trying to imagine what the First People must have felt in their long coastal and overland travels. What kind of challenges did they face moving insatiably toward the sunrise so long ago? What can we learn from their east-migrating journey? What lessons might I learn from my own ancestors who moved insatiably west across an ocean and a continent?

When we came to China Poot Bay to settle as the first white people to take up permanent residence there, we were not surprised to find evidence that the original residents built on the very same ground for the very same reasons as the original builder of the unfinished and deserted little cabin we had purchased. On any coastline in the world, if you want to identify the ancient archaeological sites, you need simply go to the places where the first colonists settled. Arriving before there were any modern constraints of property ownership, the new arrivals simply settled in the best place. They sought shelter from storms and north winds, proximity to food sources and drinking water, and natural materials for building shelters. Our cabin builder thus located his construction directly on top of an ancient archaeological site.

For anthropological purposes, we might compare the bay where we live and its inhabitants to the Middle East where, over the ages, civilizations and cultures waxed and waned, came and went, flourished and vacated. Eskimo, Aleut, Sugpiaq, and Athabascan, such as the De'naina, took their turns occupying the area and left almost no evidence of their passing. Since none of these cultures had a written language, we know little of their lifeways. Thank goodness for a succession of tough, adventurous, and inquiring ethnographers and museum collectors who had the temerity to record what little was left of the myths, legends, and customs of the people before they were lost.

Elsewhere around the world, indigenous peoples in tribes and villages had imploded or disappeared entirely under the pressure of the overlay of a newly arriving people with advanced technologies and radically different cultural ways. While wave after wave of Alaskan Natives died from introduced diseases such as measles and influenza, those who survived remained surrounded by wildlands, and some maintained the deep roots of attachment to subsistence ways. We had the impression that the remnants of these First People had a chance of holding on to what was left of their culture in the face of the modern world. When Diane and I visited in the homes and villages of the Native people who befriended us, we came as visitors from a modern world by and large unknown to them. The members of our tribe had long been removed from regular contact with wild nature. Our fellow Americans are today largely unaware of the essential services provided by and the cultural importance of wildlands, making it more abstract and difficult for them to define and respect such places and the peoples who occupy them. Perhaps those out of contact with wild nature are more compromised by the separation from the natural world than those nurtured and defined by it.

Turning the clock back in China Poot Bay to the earliest times identified by radiocarbon dating requires a long look back into the history we are familiar with. Two thousand years ago takes us back to Christ and Buddha, Mohammed and Lao-tzu. Four thousand years ago predates even the predecessors to the more modern Egyptian pyramids. Six thousand years ago predates the first written language in the Fertile Crescent of the Tigris and Euphrates, where modern Western civilization began. For most of this period, the First People lived and hunted, sang and danced, and yes, told stories on the very ground where we now live, where our children were raised, and where they are now raising their own children.

We modern, Eurocentric, recent arrivals to Alaska make much of those early white settlers that we call pioneers. We tell and retell their stories, making them our own. We take pride in numbering them among our friends; we name our streets and our children after them. It is valuable of course to remember and honor our own ancestors, their traditions and legends, but to do so without honoring the First People is to miss an important element in being whole in this far northern and far western land.

In 2005, I had the good fortune as a senior advisor to facilitate a gathering of twelve hundred delegates from sixty nations who gathered for two weeks in Anchorage, Alaska, to focus attention on what remains of the world's wilderness. The organizers, the staff, and the board of The WILD Foundation sought to involve the Native people of Alaska and invited one of the most prominent leaders of that group to be an honorary chair. Byron Malott, of Tlingit heritage raised in Yakutat, is equally at home in Washington, D.C., in a Brooks Brothers suit as he is in his role as native leader. As former executive director of the Alaska Permanent Fund, he oversaw management of billions in investments. Banging loudly on the podium in front of the international audience, he reproached us: "You white guys got no right dividing up our land with your white words, calling some of it wilderness. We don't have such a word; all the land is the same to us. This has been our home for ten thousand years and we don't want you to put your special names on it, talking about it as if you own it."

We knew that the ancient people who lived on our home ground had crossed the inlet — and even the wider and more dangerous passages between the Kenai Peninsula and Kodiak — in skin-hulled kayaks and multiperson bidarkas.

Europeans had been making contact with the indigenous people of Alaska since Vitus Bering, the first Spaniards under Bodega y Quadra and the English under James Cook, who touched Alaska in the middle and late 1700s. Diane and I saw ourselves in a closing continuum or perhaps as an end point of those historic interactions. To find our own places in this spectrum, we went out of our way time and again to make personal contact and create friendly interaction with the people of the land. As a young couple and new to Alaska we had the great privilege to get to know personally many Native people in their villages, where their closeness to the natural world and subsistence skills set them apart from most other people in the world. We did not entirely understand one another's motives. It was as if we were from a different planet, but we proceeded on trust to build what would, over time, become solid friendships.

The richness of archaeological material close to our cabin site and scattered all around the shoreline of the estuary speaks to us every day of the activity of Native people over the course of a very long time.

Just a short walk to the west is a series of caves whose floors are littered with the debris from centuries of occupants. After thousands of years of use, this very spot probably fell silent not long after the first Russians arrived on the Kenai Peninsula and conducted a raid, perhaps on this very spot. They killed the men and boys and took the women and children into slavery.

There were others out on the land and sea during those times collecting for the world's great museums. Before I took my ten-year-old son, Morgan, to Nunivak Island in the Bering Sea to go to school with the Eskimo children as part of his homeschool education, I read about Edward Nelson's time there in 1878. He was collecting for the Smithsonian Institution in Washington, and they called him "the man who buys good-for-nothing things." Nelson was especially interested in the fascinating masks of the Bering Strait region that played a variety of roles in the culture as dance and performance paraphernalia or as funerary objects. After use, these "good-for-nothing things" were often left on the tundra to return to the elements. Some of these artistically remarkable masks anchor whole exhibits at the Smithsonian in Washington, D.C., and at the Smithsonian's stunning new Arctic Sciences Center in Anchorage. Purchased originally for next to nothing and sold today at Sotheby's, one of these items would easily bring tens or hundreds of thousands of dollars. Nelson traveled widely, collected diligently, and wrote extensively about his observations, thus saving for us — and for the Native people of Alaska themselves — priceless objects, timeless stories, and insights into a culture now largely vanished.

The people were so intimately involved with every aspect of what surrounded them in the natural world that of course their myths, legends, and even their daily behavior were interwoven woof and warp with an intimate knowledge of nature. They doubtless knew how every single part of a living thing tasted and how it might be used. Some have speculated that the Unangan of the Islands of the Four Mountains in the Aleutians perfected mummification in part because of their handling of sea otters, whose bodies, whose muscles, tendons, and ligaments, are so much like our own.

My beautiful friend Terry Rothcar told me that her real name is Chaas' Kwoowu Tlaa, "Mother of Salmon." It is an honorific name passed down from her ancient heritage, and it denotes her clan lineage. Honoring her ancestors, she has rebirthed the ancient art of spruce-root basketmaking and advanced it to an artistic pinnacle of international acclaim. She digs for the roots in the dirt with her hands. "The ancient ways of gathering," she said, "are a rich lesson in today's world." Across the continent, Robert Frost also told us about "The Need of Being Versed in Country Things" in a modern story that sings of this significance but in a language more comfortable to those tied to Europe.

The recognition that there is special wisdom in indigenous knowledge is linked to the growing awareness that there is intelligence everywhere in nature. This truth is finally coming full circle and completely changing our view of the world and our place in it. We no longer dismiss myth and legend as fanciful but have begun to realize that they contain important facts and timeless truths. We are coming to the realization that people lived on this green earth for a very long time without harming it. We can and must use the combination of technology and traditional knowledge to attempt to restore some of what we have lost. And although much has been lost forever and simply will not come again, these are the most exciting of times with new and startling discoveries going on all around us. Perhaps never before in human history have individuals been better able to shape the world's perception of itself.

Suddenly we realize that it is all integrated, us and nature, us and wilderness, us and slime molds and elephants, us and bananas and monkeys and feather duster worms at the bottom of the oceans' boiling vents — we are all the same. *Gaia* is one word used in the modern parlance. We are seamlessly, magically unified into one magnificent entity, and the harm we do to any we do to all, including ourselves. There is not a minute to lose in communicating this to the greatest number of people possible.

The people who collected objects and saved their stories recorded that for the First People, every tree and stone, mountain and river, fish and animal had not only a name but also a spirit and a rich tapestry of myths, legends, and stories as compelling as the stories of Odysseus or Ibn Battuta. Across Cook Inlet, I made a point to visit a significant place in regional Native history. There was a great stone on a ridge between two watersheds. The stone sat in the exact place where a bladder skin of water poured on the ground to the west of it would eventually run to the Bering Sea; water poured to the east of it would run to the Gulf of Alaska. The hemispheric significance of this spot could not have been adequately recognized until modern times when navigation and maps gave us grand-scale perspectives. But the fact is that over the course of thousands of years when the people traversed that route between the two watersheds, they made offerings of food to that rock as if it were a person. They had conversations with it and treated it with great respect.

The people of Kachemak Bay are fortunate to know Frederica de Laguna's book *The Archaeology of Cook Inlet, Alaska*. She spent time on Yukon Island and elsewhere around the bay. In addition to her landmark archaeological work, she gave us two delightful books of historical fiction. *Fog on the Mountain* has an important historical parallel in Mardy Murie's book *Island Between*, set in Yemegaq's village. Both authors drew on their comprehensive understanding of the prehistoric peoples of their respective sites. Both lived among the people in the very villages discussed in their writing. Freddy sought to supplement her bare-bones student budget by writing for popular publications while living in Seldovia in Kachemak Bay. Mardy wrote to supplement their meager income when she and Olaus lived in the village of Gambell on St. Lawrence Island in the Bering Strait. Their writings are lyrical, interesting, and fun and address a part of the population that would not wade through the more academic publications dealing with the same places and people.

One of Freddy's stories begins, "There was once a village there." This has been a traditional way of opening stories around the world since time out of mind. In the famous book and movie *Out of Africa*, Isak Dinesen says with great emotional authority, "I had a farm in Africa, at the foot of the Ngong Hills." I clearly remember my goose bumps when she uttered those words, because I resonated with the fact that I had such a place, a home that I loved and to which I felt I belonged. No matter how fantastic the legend or supernatural the myth, such stories were always anchored to reality by the assignment of its location to some particular spot known to the listeners. Every rock, every point, every sandbar or gravel beach seemed to have some story handed down from the people who were first connected with it. They told us that the shores of Kachemak Bay was dotted with the sites of legendary settlements, and "one could never tell in advance whether the village would house people or perhaps the manlike souls of mice or bears, beavers or codfish."

The ancient name for our bay was Tsayehq'at, and it is thought to mean "a place with caves." Since there are cave paintings in Peterson

Bay and Sadie Cove just to the north and south of us, it seems reasonable that there would be caves with paintings in cave-studded China Poot Bay as well. In addition, there is archaeological material to be seen in many of the caves on our home ground. We even found a great dry-floored concavity with an underground room. One wall is the solid rock of the cliff face, and the other three walls are carefully laid flat square stone. The roof was made of heavy timbers with flat stones and earth on top. When I revealed this hidden place to my anthropologist friends Janet Klein and Peter Zollars, Peter and I were barely able to fit inside. There was barely headroom in what might have been a seal hunter's shelter. Inside was a hand-carved board with a square hole, perhaps similar to what Johan Adrian Jacobsen, another early ethnographic collector, found in this area in 1883 and described in his book *Alaskan Voyage 1881–1883: An Expedition to the Northwest Coast of America*. He traveled by skin boat from Kodiak and up the coast after an overnight on Yukon Island, "where [he] bought several lamps and dance rattles" for the Royal Ethnographic Museum of Berlin.

One year, an interesting German woman from Berlin fortuitously visited the lodge, and I called her attention to Jacobsen's book. She was surprise by the fact that he collected for the Royal Ethnographic Museum in her hometown. I read to her from his accounts as we stood in the exact places he described "at the foot of the third glacier." At my request she returned to Berlin and made a beeline for the museum, where buried in the dusty basement were the long-forgotten collections from Kachemak Bay. They had survived two world wars and the devastations of that city by being taken by the curators to the countryside for safekeeping. When our team member Janet Klein, local anthropologist and author, visited Berlin with support from Homer's Pratt Museum and the American Society for the Humanities, she was able to repatriate some of the collection to Alaska for closer examination. The new exhibits at the Smithsonian's Anchorage Museum likewise present for our spellbound gaze long-hidden works of craft from the basement of America's Museum on the Mall in Washington, D.C., which are as fascinating and ingenious in their own right as the treasures of Tutankhamen.

I have a special fondness for the four creeks that feed the south side of our estuary. The birds and furred animals share this attraction, as do the salmon when their season comes. I made a beeline for a favorite bend in a stream where the gravel in the clear water speaks a certain language to me that I do not understand but appreciate; it had been calling my name. I recalled the smooth stones looking so comfortable there on the bottom not far from the eelgrass beds. It reminded me of one of my favorite legends of the Aleutian people wherein the chief's daughter, having been carried by murres to the mouth of a cave high on a cliff, "watched how the whales enjoyed breathing." Surely, I thought, the fish and animals and the ancient people who knew this place must have enjoyed just as I do looking into the moving water where it has cut the bank and rounded the colorful stones.

Later I shared what I had seen with Janet Klein. The next day I showed her the otter trail, and some months later she returned with permission to dig and recorded an ancient site that was radio-carbon dated to sixty-two hundred years before the present. In her respectful way, she asked permission and reported her finds to the local Native group. These First People in our area were the ancient ones whom archaeologists describe as members of the Arctic Small Tool Tradition.

Elder Peter Kalifornsky — one of the last speakers of the De'naina Athabascan language, the lingua franca of this area — and I were talking one sunny day. I told him about my years hunting and commercial set netting for salmon around Iliamna, both the lake and the mountain.

"What does *Iliamna* mean, Peter?" I asked. He smiled with a chuckle, and the wrinkles at the corners of his eyes deepened. A far-off look overtook him. Taking a deep breath, considering perhaps a simpler time, he answered softly, "*Iliamna* — that means a good place to live." There was a twinkle in those ageless eyes.

Having spent his boyhood raised by his uncle in the old ways on what today we call Polly Creek just north of Tuxedni Bay across Cook Inlet, perhaps he was injecting his own nostalgic emotions, or maybe, just maybe, that is what his people had called it for thousands of years.

The first Native people encountered a world that was fresh, pure, and clean. It was handsomely populated with animals. Those first white people encountered a magical, almost untouched wilderness. Each of us has the opportunity and the responsibility to tread lightly, to cherish and honor the land, to maintain clean, unpolluted, and beautiful water and air for all children who will inherit what we leave behind.

In 1967, Diane and I had not been married long when we flew from Anchorage to far western Alaska in a Piper Super Cub on floats to the watershed of the Bering Sea coast. We spent a summer season there living with two very traditional Native people, Simeon Moxi and Ester Etukemaldra, whose lineage was that of Bering Sea Eskimo. Neither Ester nor Simeon had ever left the immediate area of their home, the tiny village of Aleknagik located downriver and then down from the lake, twenty miles or so from our isolated little outpost. Their people were a subsistence-oriented group, intimately tied to the land and sea, who had lived prosperously on that coast since they migrated east across the Bering Land Bridge from Siberia. The impression that these good people made on us lingers still.

Simeon was barely more than five feet tall, with a face as deeply wrinkled as the Wood River Mountains themselves. His skin was as brown as the coffee he loved. Simeon was not his Eskimo name; unfortunately, I never learned what it was. If I had inquired and then used his real name, it would have honored him. Chances are it was the same name as one of his ancestors. His given surname, probably conferred by the owners of the trading post, was that of a soft drink popular before his birth around the turn of the century. Older than my father, he was among the last of that generation that bridged the time between those who lived in the old way in remote Bering Sea villages and the new era, when people like me came into the country with floatplanes bringing catch-and-release sport fishermen.

These descendants of the Bering Land Bridge people had worked out a balance with the land and sea over many thousands of years and saw the fish as a different kind of people: they had spirits that should be propitiated and not offended. Being right with the fish, returning their unused bones and skins back to the water assured their annual return. Honoring this tradition even today, I make a point to return any unconsumed fish parts back to the sea.

"Them fish, they want their clothes back. You give 'em back, then they come back plenty."

What could we learn from these people? What truths were before us to be discovered about life itself and our place in it? What lessons could we take back to our own little cabin in the woods by the estuary and incorporate into our own lives and ways of living on the land? What might be drawn out of ourselves in the process of addressing these questions? Perhaps above all, our interactions with these tough and enterprising survivors buoyed our optimism that we too could one day count ourselves among them as survivors in this northern place. It was important of course to have humility, timidity, and even fear of the many unknowns, but we needed confidence and self-assuredness, or we would never be able address the challenges ahead. We looked to the people, to Simeon and Ester, who didn't say much, to the land and the animals as our teachers.

There would be many voices to heed, and we needed to be able to listen and learn from the wind, from the stars, from the water sounds. The Native people said that there was a time when animals and people spoke the same language. Perhaps that time was not so very long ago. Listening carefully on moonless nights, we thought we might even hear its distant echo.

From time to time even today I fantasize about returning, not to see if our little log cabin still stands beside the Agulowak River, but to the Native village hill of Aleknagik at the end of the lake. I would inquire of an elder to learn where Simeon was buried. I would ask relatives for permission to make an offering of food at his grave site. A can of Spam and some Pilot Bread would be just about right. Smoked fish would be even better. My mind's eye says that the relatives would lead me quietly to a place where there would be no white man's burial stone, just the fireweed and twinflower, sphagnum and crowberry, and just maybe, whiskered Ketok, the wise old raven. I would like to believe that he would be there looking down from a black limb on a white birch. After Simeon's descendants left, I would have a last visit with my old friend. Making sure that no one but the raven was looking, I would leave as a gift my favorite Buck 505 pocketknife that I would push deep into the moist tundra.

A brown bear looks on at Chenik Bear Photography Camp. Photo © Boyd Norton.

CHAPTER 7

Chenik Bear Camp

"This is how it must have looked when the world was
new, no trees, living waters all around."

—Vuzamazulu Credo Mutwa,
a Zulu Sangoma and Shaman,
on being shown photos of Chenik in Kamishak Bay

If I could live a dozen lifetimes, I might trade them all for that spring day as I probed the mouth of a sizable but nameless stream in Akumwarvik Bay. I was alone in the open seventeen-foot Boston Whaler, and exploration was my game. I had no particular reason or need to be in the shallow reef-infested area, where I motored slowly along, stunned by the clarity of the water and the absolute stillness of the air and sea. Dark eldritch shapes loomed around the shadowed bases of scattered small islands, whose sheer concave faces defied the lust of a skilled climber. In such a place, one monitors a tide book again and again as nervously as a doctor monitors the pulse of a child. A small mistake could mean that the boat would go dry and not float again for two weeks on the next lunar pulse. The weather trend that I had been watching overhead told me that staying the night wasn't too risky. A welcoming beach bid me "come," and I made a basic camp among the driftwood logs and giant bear tracks. In what many would have found a menacing environment, I fell asleep in the sunny blond sand and had not been at such peace since I was a child in my father's arms.

The Kachemak Bay Wilderness Lodge was in its seventh year of operation in 1977, and we foresaw the urbanization of the Homer watershed in the years ahead. Looking beyond the horizon to where the sun sets, we sought a place where our clients would always have the opportunity to experience the kind of real wilderness that we had come to love in the 1960s. The few old-timers who knew anything at all about the land south of Ursus Cove and Bruin Bay spoke with such awe that we were drawn to explore its forbidding coastline.

I began going into that Kamishak country in 1972 and started looking for land on which to establish a camp, with emphasis on brown bear observation. In 1977, we drove the difficult five-hundred-mile round-trip to Anchorage in the dead of winter, where I made a visit to the Bureau of Land Management (BLM) offices to check on the status of a piece of land, about a hundred miles west of the lodge across the tumultuous Lower Cook Inlet, that had caught my eye. Little did I know then, but the possession of the pink sheet of paper with registration number Ak016-8064 would become critically important in the future. I paid the fee and filled out the forms. Where the purpose of the camp was called for, I described the scientific, educational, and training purposes we envisioned. Next we hired Alex Flyum, the same Bering Sea Eskimo who had taken us to China Poot Bay the first time, to take us to Chenik Lagoon just north of the McNeil River State Game Sanctuary the following spring. The state created the McNeil River State Game Refuge as an extension to the sanctuary eleven years after our entry, which surrounded but excluded our parcel.

Looking down from space at this region, one sees geography of hemispheric proportions written in blue and green. The earth's tectonic plates collide here, and as they meet, one plate crunches up and over the other, pushing the peaks of vast mountain ranges up and even through the clouds. The low, windswept pass where the camp sits calmly divides a mountain range that begins with Denali to the north. Moving majestically south, it begins a thousand-mile arc to the west as it sinks into the Pacific abyss, leaving the tops of those same mountains sticking up out of the sea to become the Aleutian Islands. Almost two score of cinder-spitting volcanoes stab through this restless collision zone, and three of the stratovolcano troublemakers can be seen from the camp.

The view north from the camp encompasses the near terminus of the ten-thousand-mile sweep of the Pacific Rim and Ring of Fire. Seen from the front steps of the little cabin to the north is a two-mile-high volcano whose lofty peaks are draped in hanging glaciers. Chenik is in a land of fire and ice, brown bears and moose, whales and octopi, orchids and frogs. From mammoths to moths, forces both vertebrate and invertebrate have shaped this land as ice ages have come and gone. It might be dismissed as too wild and savage to be of use to anyone, were it not for the remarkable fecundity of its brief summer. An artist's palate of rowdy colors splashes across these seaside meadows. Beginning in midsummer, the magenta carpet of *Epilobium angustifolium* — or fireweed — rolls up the shoulders of the mountains in glorious display. Botanists at the camp have enumerated more than a hundred species of wildflowers, including boisterous classrooms of the eye-catching *Cypripedium guttatum* — spotted lady slipper orchids. Those flower-splashed meadows cozy up to a salmon river that yearly becomes so choked with the muscular fish that it seems there is more silver protein than water. Salmon have been here since this flow came down from the river of stars in the Andromeda. The bears apparently came from a nearby place and have been here just as long.

Chenik's star began to rise on the western horizon at a time in the state's history when only a few people had ever taken a close-up photo of a brown/grizzly bear standing in a white-water river with a salmon in its mouth. I knew that we could accomplish this for our clients one hundred miles west of the lodge, in a little-known place on the little-known Kamishak Bay coast. I knew that Chenik Creek drained a lake of that name and came tumbling out of its winding course through alder thickets, where it fanned out across a rocky outcropping. This place offered a sheltered cove where a boat could be safely moored and where bears could snap at airborne salmon, while photographers snapped shutters to their hearts' delight.

Rising above the tumult of the modern world, the great coastal brown/grizzly bears have followed a thousand miles of shoreline looking for spawned-out and washed-up salmon, seals, and whales since the beginning of time. Chenik Head and nearby Amakdedori Beach seemed to be magnets for drifting whale carcasses, and when they grounded about once per year, the bears swarmed like bees to honey.

Our camp was in the midst of the bears' well-worn trails, and the camp's philosophy was to protect and preserve the bears and the natural world that supported them. The depth of affection that our family developed for this place is simply impossible to describe. Those who worked with us and put their energies into creating the camp and those who came to know it as visitors were equally enamored of the remarkable power of the little cove that teetered between the primordial and the modern.

For the countless days that *Homo sapiens* spend tromping around in the bush, interrupting a sow bear nursing cubs, disturbing a bear chewing on a moose carcass or having a nap in the sunshine, it is a wonder that there are not more attacks and fatalities from *Ursus arctos* and his ebony-furred cousin, *Ursus americanus*. It just goes to show that bears are actually pretty tolerant of the bipeds and by and large will go out of their way to avoid a confrontation with a human.

In 1978, we carried a year's supplies of materials and a season's supply of groceries. We unloaded at night on the high-tide peak and awoke to find a smiling brown bear in our food supplies. This was not a good start because we had just created a real problem for ourselves: a bear that associated us with a five-pound bag of oatmeal. The flip side was a questionable celebration that we had indeed landed in an area thick with brown bears. The future of the camp looked bright. To celebrate, we cobbled together a sauna, using six sheets of plywood, and made a stove out of a five-gallon blaze can and the chimney from three more cans that we flattened and curled into cylinders. That night the bears in the sauna thanked the bears in the alders for welcoming our arrival.

The modus operandi of our undertaking at Chenik capitalized on a point, counterpoint at play in the environmental consciousness at the end of the millennium. On one hand, the people wanted to be up close and personal with the great bears and were willing to travel great distances, spend considerable sums, sleep in wall tents, and suffer the primitiveness of outhouses — all for the rare privilege of being in such a place with its magnificent animals. They spent untold numbers of hours just watching spellbound at the unfolding dramas of nature. The savage and sublime balanced timelessly between the mountains and the sea. The great carnivores were of course center stage and the salmon the glue that held them to this particular place, but there were wolves and wolverines, foxes and moose, and on occasion the great brown-beamed caribou bulls wandering through, who thought the camp a curiosity.

On the other hand, "Le sauvage contretemps avec le gentil," the savage and the gentle were blended together here. Our little camp was an oasis of comfort and civilization in the midst of a vast wilderness. The place and its comforts generated a rare kind of conviviality. The stream drained the tundra head and poured it through the *Elymus maritimus*, beach rye grass, past the little trout finning there at high tide's edge, and its murmuring voice brought out the best in those who beheld the sweeping view up the coast to the north, where the surf incessantly pounded the curving half-moon shore.

All the while, the great carnivores let us know about their interest in and affection for our presence by clawing the logs or actually biting on the building. It didn't much matter whether we were actually inside. "Anything good to eat in there?" was the guttural ursine language, and we understood exactly what they were saying. We shivered and shook just a few feet away from those great salivating teeth. The sound of the huge curving claws as they raked across the logs near our bed was not quite a lullaby sung by a loving mother. We were fastidious with garbage and fish waste to be sure that the bears never got any human-oriented food. In spite of that, there was often a frightening commotion in the middle of the night just beyond the door.

Since we had risked so much to make the camp a reality, we were always hungry for people to say good things about their experiences at Chenik. The cards and letters of thanks and appreciation were like water on a thirsty plant, but nothing seemed more wonderful than seeing praise in published materials like newspapers and magazines. One of the best-known photojournalist couples during that era was Peggy and Erwin Bauer, and they were among our first clients. They wrote in *Outdoor Life* magazine, "We've never experienced absolute wilderness anywhere in the world like what we saw and felt at Chenik, not in Wyoming or Antarctica nor in the furthest reaches of Hudson Bay." On their heels came a photographer with writer John Hemminway, on assignment for *Connoisseur Magazine*, and again we got a home-run story. Humorously, the feature was between an article on a British sports car called a Morgan, our son's name, and another about a distinguished couple living in a French castle. These articles were followed by others that helped us enormously, as they brought remarkable people to our little outpost from the four corners of the world.

In 1978, the first year of our operation eight miles north of the McNeil River Brown Bear Sanctuary, we had a group of ten clients from New York City. So complete was our bonding and so happy were they with their experience that some of them were still coming back twenty years later, and now, almost forty years later, we still receive in the Christmas mail the can of real Vermont maple syrup from one of that group. That was the same year that a guest on assignment from *The New York Times* wrote a Sunday Travel Feature about us. From that lucky boost came another Sunday Travel article in the *Times*, and from that came a long stream of distinguished guests from the Northeast, primarily New York City. Chenik, a forgotten little pinprick on a vast map of wilderness, ended up with bedfellows that no one could have envisioned.

From the first day we set foot on land at Chenik, we were nervously aware that we were walking on ground that had recently been trodden by the big paws of a heavily clawed bear. We encountered bear scat as often as a beachcomber encounters seaweed, usually noting from the bore the relative size of the bear, what it had been eating, and how fresh it was. Based on our assessment, we might increase our periodic clapping of hands and the requisite loud callout, "Hey bear, hey bear," even on the way to the outhouse. These forms of communication are intended to let any nearby bears know that you are in its territory, lest you come on it by surprise. Our hiking trails were bear trails, and in some places the dinner-plate tracks had, over the centuries, worn potholes in the tundra so firm and deep that they were too awkward for people to walk in.

One night when Diane was sleeping in the wall tent that we called "The Cub," she heard the telltale footfalls of a bear coming down the gravel trail that passed within a few feet of the flimsy, if handsome, structure. She held her breath as she sensed the bear pause on the trail just a reach away. As she lay frozen, the bear stretched its neck and pushed the center of her back through the tent fabric. Not the kind of event to report to Diane's parents but indicative of the rapport we had established with the great bears who knew somehow that they had nothing to fear from us — even though we feared the fact that we were but visitors on their ancient ground.

Although we have been very close to bears many, many times, only once did we feel threatened. We were packing moose meat after a successful hunt thirty miles west of the camp when a lactating female grizzly charged us. She was momentarily out of her mind with rage and communicated that to us in no uncertain terms. A bear has the remarkable capacity to come to a full foaming froth at the lips in about the length of time it takes to read this line. We were heavily loaded with the meat, and the blood was dripping out of the packs and down our backs and legs.

Diane and I had been to and from the kill site a number of times. Over the years, I have found that if the animal is carefully boned out and all the meat taken from the carcass, it takes about eleven round-trips to carry in the meat. With Diane carrying about thirty pounds and me about eighty, it would take a few more trips than when my brother and I challenged each other with how much we can bring in. The packer lies down on the ground and attaches himself to the pack. His brother rolls the pack over on him. The packer tries to stand up, first to hands and knees and with a groan, as his partner struggles to help him to his feet. The straps are cinched down at shoulders and waist until the bearer feels like a sausage, and the grueling march begins. Heaven help those who in their excessive zeal have shot a moose too far from the means of getting it home — in our case, a floatplane. There may be dense, tangled thickets, swamps of brown mush, streams to cross with the associated banks, or clefts to cross. It may be raining, snowing, or blowing, or it may be a potpourri of all of these, combining to create a bug-swarming, sweating misery the likes of which you would not wish on your worst enemy.

We were somewhere in the middle of the meat haul and struggling through very dense and impossibly tangled alder brush. Although we had traversed this route a few times before, it was near impossible to take the same route twice. Usually we take a roll of surveyor's tape, and after a passage or two when we have found the easiest route, mark it and follow the flags for the remaining trips. We stumbled and swore our way through the morass, which was generously punctuated with the evil devil's club thorns. We came to a small clearing, and at its low summit the grizzly burst out of the thicket and was on us before we could react in any way. When a bear is screaming at you from only a few feet away and popping those terrible foaming fangs, there is the uncontrollable relaxation of the sphincter muscle associated with that primal lizard-brain reaction to what looks like certain death. Fortunately, it wasn't. Having frightened us nearly to death, the sow did what sows usually do in such situations: she disappeared back into the brush.

As she departed the scene, we heard her say quite clearly to the cubs, "Scared the dickens out them, didn't I?"

We were using a twenty-one-foot Boston Whaler to support our camp activities and had worked out weekly air support from our old friends Bill and Barbara DeCreeft and their Kachemak Air Service in Homer. Clients began to come from all over the world. And at the time, that iconic photo of the bear in a white-water river with a salmon in its mouth remained seldom seen and hard to get. Famous photographers such as Galen Rowell, Boyd Norton, Erwin and Peggy Bauer, Tom Mangelsen, Art Wolfe, and others brought groups for workshops. International figures like the famous Bernhard Grzimek from the Frankfurt Zoological Society brought groups from other countries, and still others came from as far away as Africa. The conversations around the dining table were truly remarkable. The camp was humming with activity that included a wide range of scientific support activities and internship programs for young people, as we transitioned to a full-time 501(c)(3) nonprofit scientific and educational corporation.

There is, however, a grim reality to living in bear country. We still carry with us the grief and horror we felt when we learned that our treasured friend Mishio Hoshino had been killed and eaten by a brown bear on the Kamchatka Peninsula in Siberia. We had slept in his Fairbanks log cabin not long before when testifying to the State Game Board, trying to protect the bears. Being eaten by a bear, anyone's scariest thought, is a very rare phenomenon in Alaska, and such images linger in a way that car crashes or falling off of a stepladder in your living room do not. Although a handful of people are mauled by brown bears each year, the fact is that a person has an extremely small chance of a negative encounter with a bear. These odds shift a bit between urban dwellers who make periodic trips into the bush to fish, camp, or hunt versus people like us who live most of the year surrounded by bears. Perhaps like people anywhere living in close proximity with any animal that might kill you, one assesses the risks, compares them to other risks, and presses on with the business at hand, trying all the while to minimize the exposure. One does not stop swimming in the ocean because there are sharks out there.

Only heaven could have helped the small businessperson who had to navigate the largely uncharted, sometimes dangerous, and usually rancorous waters that flowed between the state and federal government and Native corporate interests during Chenik's lifetime.

During the two decades of our occupancy in this wild place, there was every reason to believe that the permit we had paid for would put us in a position to eventually lease the land from the BLM. This proved to be true when, twenty years later, in 1998, we were offered a fifty-five-year lease by the US government. This was in an early statehood era when hundreds of people were proving up on "free" land offered by the federal and state governments, intending to encourage colonists to spread out and establish homes, businesses, and enterprises to diversify and expand the state's economy.

When I went to renew our BLM permit, I was told to seek permission from the Native Corporation to continue our operation. They had made their land entitlement selection of those same lands from the federal government as part of the 1971 Alaska Native Land Claims Settlement Act. We acquired that written permission and maintained an active interface with the Seldovia Native Corporation, continuing a twenty-year-long association between the BLM and the corporation lands that surrounded, but excluded, our own. Interacting with those shifting and confused bureaucracies, I thought it wise to carefully document every visit to each office. I saved every letter and phone conversation note until there were sixty-three pounds of files, weighed on the bathroom scale. The camp would eventually make the front pages of papers across the state because of this confused three-way-claim ownership between the Native Corporation and the state and federal governments.

The corporation wrote that as soon as the land transfers from the BLM were completed, they would give us a lease. During this period, we made regular visits to the BLM and the state land offices. In spite of often conflicting instructions, the Native voice was loud and clear: "That is our land, has always been our land, and neither the federal nor state people can tell you what you can or can't do there." The Native Corporation confidently spent the better part of a million dollars asserting their claim to lands used by their ancestors before finally giving up on their claim to Chenik in 1999.

A telling written memo from the BLM stated that there was mass confusion and that we were given "conflicting information by BLM and the State." In addition, they wrote, early records of our interactions were destroyed by the BLM. In a 1996 comprehensive case review at the Department of Fish and Game, a letter stated that they did not find evidence that the McBrides had been negligent and therefore recommended that they receive the BLM lease. That comprehensive review of the history and its documentation in order to understand it fully would be the equivalent of undertaking a PhD dissertation.

The ultimate demise of the camp, and the political and cultural events leading up to its hostile overthrow, constitute a story perhaps archetypal of frontiers throughout history when ancient ways in ancient places collide with colonial attitudes, and those in turn merge with or crash into modern sensibilities. It is interesting to observe that the most remarkable populations of wild animals in the world often end up at the epicenters of widespread attention: witness Dian Fossey and the gorillas in Rwanda and the ongoing efforts of Jane Goodall to preserve the Gombe forest in order to protect the chimpanzees there. The story of George Schaller and Peter Matthiessen's work to save the snow leopards in Nepal and Tibet is paralleled with my fellow African game rangers Ian Player and Paul Dutton's heroic efforts to save African rhinoceroses and dugongs. As I write, one of my closest confidants, Boyd Norton, is fighting to prevent the building of a road across the migratory path of the great herds of Tanzania and Kenya. There are numberless and nameless people all around us who are working their hearts out to protect our natural heritage. What strange quirk is it that allows people to harm nature in ways that are so obviously detrimental to biodiversity and to show such disrespect for the land and sea? How do they fail to see that such disrespect is, after all, disrespect for themselves?

We hemorrhaged money for years on lawyers, caretakers, aircraft charters, and plane tickets to Anchorage and a thousand miles farther on to the capital in Juneau. We met with decision makers to point out the unfairness and the associated constitutional illegalities of our forced eviction.

The final evacuation of the Chenik Brown Bear Camp now at hand had been a long time coming. For years, its fate had hung suspended like a pendulum. Time and again the scale had teetered this way and then that, but now it was settled: the long struggle to maintain the facility had been lost in rancorous bureaucratic arm wrestling between the state and federal governments and a Native Corporation. The McBrides' formally permitted tenure had ended, and the camp would be left to an uncertain future.

This part of Alaska's wild coast is washed by some of the world's largest tides. Our regretful evacuation of the Chenik Brown Bear Photography Camp could only be accomplished during a seasonal cycle of extreme high tides, otherwise the landing barge could not enter the shallow intertidal lagoon. During the twice-daily low-tide cycle, the ocean briefly gives up its incessant gnawing on the shoreline and retreats to expose tens of thousands of acres of glorious eelgrass flats beside the camp. This is feasting ground for skies full of migrating waterfowl and wading birds in spring and fall. Oystercatchers and ground-nesting bald eagles, Pacific eider ducks and water ouzels are but a few of the hundred or so members of the avian community who pass through Chenik Head or call it home.

The arrival of the landing craft and the throaty chugging of its big diesel engine had broken the absorbing silence and isolation of this wild place. Its low-frequency vibrations could be felt in the cabin. For days now there had been a flurry of activity, as a lively generation of occupancy in this remote outpost was being drawn to a close. Sad and even tragic though this ending was, there was no time for maudlin emotions; there was work to do.

There had been a time when that woodstove was the most important thing in the world to us. It gave us heat when we were cold, it cooked our food, and it became the very axis of our lives as newlyweds in those first challenging years in the little log cabin beside the China Poot Bay estuary. When the babies arrived they sat on top of that stove in a galvanized tub having a fine bath. Diane mastered the art of producing the finest bread in the north. Those hearty loaves and the warm baby bottoms grounded us to the earth and each other as perhaps nothing else could. The stove had assumed many of the qualities of a person, and we loved it almost as dearly.

Twenty-five years after we built our first fire in that little fire-box, the woodstove's very life hung on a simple decision, but it was not forthcoming — there were too many other things to think about. The room had been emptied and the stove sat alone within the four bare walls. The people were outside, below the cabin on the shore, absorbed with the sad challenges of vacating this remote place they had occupied for more than two decades. The basic twenty-four-by-thirty-six-foot frame cabin at Chenik faced a protected intertidal lagoon to the north. To the west was a small stream that had been named for our son, Morgan, and its murmuring voice could be heard throughout the room.

There were fifty-five-gallon drums of unused fuel, solar panels, furniture, propane stoves, washing machines, mattresses, chain saws, tools, wheelbarrows, and generators. There were cloth napkins, curtains, and wine glasses, framed original art and books, recorded tapes of Mozart and the Beatles, and the thousand and one other things used in such a far-off place to make it utilitarian yet homey. The nearest store was one hundred miles away by boat or floatplane. Scores of utility boxes filled with other accoutrements of wilderness living had been stacked at the high-tide line in anticipation of this moment.

Surrounding the cabin was coastal tundra whose broad meadows of grass and wildflowers were studded with dense patches of head-high alders. The shoreline meandered north to Amakdedori Beach, Bruin Bay, Ursus Cove, and an unmeasured one hundred miles of bear-freckled coast just to reach the nearest summer people who worked seasonally at salmon set net sites. There were no trees in this windswept place. The windows had been bolted shut with heavy bear-proof shutters, for this was brown bear country. While bears had never tried to break into the cabin as they often do elsewhere in the Alaska wilderness, the shutters had been good insurance. The cabin and its little lagoon are located on the southwest shore of Kamishak Bay in Lower Cook Inlet about 150 water miles south of Anchorage. Augustine Volcano looms on the eastern skyline an hour away by boat. The sulfurous fumes venting from the summit reveal the wind's direction and intensity like an airport windsock in the middle of Kamishak Bay here on the shoulder of the Alaska Peninsula.

During the period when the vertical change is at its greatest, the tide rises or falls an inch per minute and can vary more than thirty vertical feet in six hours. Entire cubic miles of water rush in and out of Cook Inlet, whose watershed covers thirty-three-thousand square miles. This day's tide had been identified months ago, and it was to peak at 22.2 feet in height at 4:03 PM. The busy men were racing to load the barge during the few short hours when the tide paused briefly at its peak. Our hope was to avoid the barge going dry for an additional twelve hours and thus incurring another day's charter expense. The barge charter had been estimated to cost $4,000, a significant sum considering that the camp had been unexpectedly closed and without income for four years.

Down beside Morgan Creek, strong men in coveralls were working quickly to load the landing craft with the tons of supplies that had crossed the inlet by boat and floatplane at great expense over the course of two decades. The barge had just traversed the hundred-plus miles of open ocean from the nearest town of Homer. The evacuation had been postponed again and again, but finally here it was, already late September, and the last viable weather window of opportunity was at hand. It would not come again. As the seasons begin to change and the tide cycles build in concert with them, the storms increase in severity. This part of Alaska is infamous for the savagery of its weather. Many the person, boat, and aircraft have met an untimely end at the hands of the raging winds that tear through this low pass between the Bering Sea to the west and Gulf of Alaska to the east.

That night in the moonless dark, the stars shone with the special crispness that comes to Alaska's coast in the fall. There was not a breath of wind from any point of the compass as the barge backed away from the loading spot at 3 AM. The reverse prop wash lifted pebbles and rocks from the bottom and threw them loudly into the steel hull tunnel surrounding the shaft and propeller. The skipper and I had together examined the two large props when the barge was dry. Their leading edges were rounded by scores of such shallow-water operations.

Although the sea was perfectly calm, there was none of the foreboding that often accompanies such stillness before a storm. There might just as easily have been a sizeable surge on the lagoon beach compromising the entire process. There were rocks and reefs all around our escape route, any one of which could have punched a hole in the hull. Tom Van Zanten, the veteran skipper, had been in this tricky lagoon only once before.

It was a mystery how Tom could have remembered his way around the channel from a single visit years before. The engine was barely idling. Sightless, the barge groped through obstacles both real and imagined. It was pitch-black — as black as the inside of a cow. *Bumppp, grrrind, screeech*! I imagined a sleeping grizzly waking suddenly, nose up to sniff, head slightly back, and considering if what it just heard was an animal in distress and perhaps tomorrow's dinner.

"Dammit!"

The boat ground to a stop as steel screamed fingernails-on-blackboard across solid rock. Tom cleared his throat, put the barge in reverse, increased throttle, and the same hideous sound was repeated, only backward. How to correct the course in the pitch-blackness? Which way to turn now? The night air was pure, crystalline — the shoreline was visible only as a black mass against the starlit sky. There was no depth perception or sense of scale. Mrs. Allen's stove sat at the bow waiting fearfully to be dropped in the inlet. As we felt our way slowly, carefully out of the lagoon, the northern lights came out in the northern sky and offered a luminous send-off to the sad departure.

The eighteen-hour crossing of the inlet was a slow, ponderous affair. From time to time the Global Positioning System indicator registered our speed as 3.1 knots, then 2.4 knots. You could walk that fast! From time to time, as the big white rollers smashed into the bow, the square-nosed craft nearly came to a halt. There were four bunks aft of the wheelhouse. We caved into them as fatigue prevailed. I had for some reason respectfully thrown a weighted tarp over the stove as we left the lagoon. Tons of green spray exploded against the blunt bow and crashed over the load and chased out the scuppers. The stove crouched under the salted tarp. There was no chance to lower the bow ramp and send the stove to Neptune at the bottom of the sea.

Back at last in the Homer barge harbor, the rusty loading ramp thumped down onto the sandbar and the hours of unloading began. The tide was flooding this time, so the skipper kept the engine in gear at low idle and edged up the beach a little at a time as the tide flooded. A pile of materials that would go to the dump was building on the left. My old friend Otto Kilcher crept down the ramp and slid his forklift tongues under the pallet on which the stove was perched.

"Hey, Mike, this old stove goin' to the dump or to the pile of good stuff?"

"No way — that is a great old stove. Wouldn't think of takin' it to the dump."

On January 15, 2004, Chenik was burned — a day after the state manager of the region met with her staff to plot the useful future of the camp as a scientific, educational, and training facility. One of that group, a pro-hunting, disgruntled Alaska Department of Fish and Game employee, chartered a helicopter at state expense, flew to Chenik, doused each building with kerosene, and one at a time set them alight. He watched them burn to the ground in the brisk north wind.

Those who colluded with him looked the other way, and that was the end of that. The Dodge City Wild West shoot-'em-up behaviors of 1840 were still alive and well in Alaska. Interestingly, the FBI was beginning to investigate state legislators who would in the coming years go to jail. Those who caused the burning of Chenik — let alone the one who lit the match — were unscathed by the flames. I am reminded of the deranged fellow smashing the *Pietà* in the Vatican with a sledgehammer.

For twenty-five years, Chenik offered a refuge for stranded fishermen, hunters, and aircraft pilots. The oldest admonition in Alaska is to "leave the latch string out," implying that remote cabins are traditionally stocked with survival needs. For almost a quarter-century, there had been food, stove, fuel, bedding, and survival equipment available in the sauna for those experiencing an emergency. This resource was used more than once by people in need. The closure of this chapter at Chenik will be the end of this time-honored tradition on a stretch of coast as wild and inaccessible as any on earth.

"For of all sad words of tongue or pen, the saddest are these: 'It might have been!'" Notwithstanding the romantic connotations of this phrase, it does refer to the deepest emotions a person is capable of. Using these words to capture the emotions surrounding the loss of something we held so dear helps bring the pain to a close. It is sad, but for any misfortune in life there is a new beginning. One crosses that river and moves to new ground. To have arrived at the place where I stand today without rancor or regret has been one of the greatest challenges of my life. I am finally ready to distance myself from the word *loss* in favor of a word less accented with regret. In order to maintain my own sense of wonder in the natural world, unobstructed by the more base human instincts, I have forgiven those who did this vile thing; they will have to forgive themselves.

There is much to celebrate. There are ample opportunities each day to say thank you to the earth and the sky and the sea. There are countless ways to be proactive, as together we who care about this great green-and-blue planet work to compassionately share our passion, our enthusiasm, and our commitment with those who do not see the value in doing so.

The cockpit controls of Steinway *eight-oh mike. Photo © Boyd Norton.*

CHAPTER 8

Of Flights and Waves

BARREN ISLANDS AND KAMISHAK BAY WATERS

400 PM AKST THU JAN 12 2012...

STORM WARNING THROUGH FRIDAY...

...HEAVY FREEZING SPRAY WARNING

THROUGH FRIDAY NIGHT...

TONIGHT...NW WIND 55 KT WITH GUSTS TO 70 KT.

SEAS 18 FEET HEAVY FREEZING SPRAY.

—the National Weather Service

The throttle moves smoothly forward with the pressure from my left hand, as my right hand pulls the control stick back between my legs as far as it will go. The engine roars like a startled lion as it advances to twenty-three hundred rpm, and the floatplane lunges forward across the still surface of the remote mountain lake.

Ben-Hur must have felt just like this when he sent the tip of his long black whip whistling between the ears of his trembling stallions. With that sharp crack, they are off in a lunge that pushed him backward in the chariot, just as I am pushed into the survival suit that fills the backrest of my seat.

My winged chariot charges forward in a fury of white spray. The Piper seems ready to challenge the whole world. In a moment, my own 150 lunging stallions, my 150-horsepower Lycoming, fired with 80/87 octane fuel aided by miracles of steel, chrome, and aluminum, is transformed to a waterborne thing of tight fabric, bent tubing, and Plexiglas. In far less than the length of the chariot course, I have been carried high aloft by a fabulous creature that was born to be in the air. With the trailing edge of the horizontal stabilizer held slightly up, the airspeed indicator is steady at fifty-six miles per hour. The angle of inclination teeters at about thirty-five degrees. In the space of just sixty seconds we are hundreds of feet in the air. Revealed now to our privileged eyes is a festival of the gods: mountains and glaciers, lakes and streams, and waterfalls beyond all counting. There is not a living soul for miles in any direction as we survey this grandiosity with the eye of a hawk. There is that same sense of bewilderment that smote us the first time we saw this wild lake we call Loonsong.

There is a powerful magnet that draws men and women in Alaska into the air and out over some of the world's last remaining truly wild country. Stretching as it does from east to west the distance from San Francisco to New York City, with only a few squiggled lines of highway in all that vastness, flying is clearly the only way to have a hands-on grasp of the enormity and diversity of The Last Frontier.

Sitting at the controls of your own aircraft offers a set of pleasures that are hard to describe. I am disinclined to simply call it "fun," because this word implies a carefree attitude. The pilot must always be alert, and in the words of an experienced bush pilot who trained me, "You must always be out in front of the airplane, not just in the seat. Out in front means anticipating, thinking ahead; not reacting. You always want to know where you would land if the engine quit. You always want to be well prepared for what happens next."

My son and daughter love to fly with me on camping, hiking, climbing, and berry-picking trips. Their appreciation recently found expression when they offered to buy a small single collapsible kayak for my birthday. An aircraft can land a person in wild country with a fishing rod, shotgun or rifle, berry bucket, or camera like nothing else can do. Add diving equipment, scientific collecting or sampling materials, or my collapsible kayak, and the geographic range for these aircraft-facilitated pursuits expands exponentially.

Owning and operating a floatplane for use in bush Alaska presents as many opportunities as challenges, as many advantages as liabilities. Entry into this small club is often limited by money, for it is not an inexpensive pursuit. Funding is often the limiting factor even if, as we do, you use the aircraft as a tool performing a function for business. If the funding hurdle can be overcome, then specialized skills must be acquired and kept finely tuned, which means flying regularly year after year. Even if you are not a licensed mechanic, there are time requirements for maintenance and upkeep, all of which adds up to a serious commitment at many levels. I stayed out of the cockpit for several years during the full-speed-ahead development of the lodge. I realized that I had too many balls in the air, too many distractions, to be able to fly safely. Leaving the ground, whether on a stepladder in the kitchen or on an extension ladder to a rooftop, implies that you might fall down and hurt yourself. Falling down when flying in bush Alaska as often as not means falling down where few people are close at hand to help you get up again. If you hurt yourself out there, you have a problem.

The bush pilots' fraternity that I hoped to join in 1966 as a young bachelor new to Alaska *was* equivalent to that of the Wild West cowboys. In Alaska, the fame of the pilots matched that of John Fremont and Bill Cody. Pilots Don Sheldon and Bob Reeve had books written about them, as did a dozen other Alaskans who flew beyond the ends of the few roads. Even before meeting some of these Alaskan legends, my head had been filled with the stories of Saint-Exupéry in France and over the Andes, Beryl Markham in Africa, and the Lindberghs flying the Siberian coast. Each of them spoke to me across the miles and years about a life of adventurous flying in frontier situations. I was eager to follow in their prop wash.

Commercial pilot's license or not, it was clear to me that the first several hundred hours of flying after a pilot is set free are the most dangerous and when accidents are statistically most likely to occur. One might have the skills but not the wisdom that comes only with hard-wrought experience. Some say that you don't know much about flying the bush until after your first thousand flying hours, and I reluctantly subscribe to that harsh reality that requires a stiff dose of humility in the face of high-octane ego. The number may be arbitrary, but there is no replacing the hard knocks of experience of flying in bush Alaska.

As a recently fledged commercial pilot, I scanned the broad field of flying Alaskan hunting guides to identify the one whom I thought was the most reputable and likely to be the most successful in the years ahead. Bob Curtis took me under his wing. I began guiding for Bob deep in the Alaska Range in the fall of 1967. I soon learned that his client list looked like a "Who's Who" of American business, and European nobility were frequently his guests.

When my air force captain's uniform came off in 1969, Bob gave me Super Cub 5988 Zulu, and Diane and I flew it to the Agulowak River in the Tikchik Wood River Wilderness system north of Dillingham in far west Alaska. We were hired to run a small fishing lodge for Bob and his wife, Gayle, while they built what would become one of the world's premier trophy rainbow-trout fishing destinations, the famous Tikchik Lodge, sixty miles north of our camp. Bob had bought the inholding of land and minimalist buildings on the Agulowak River from John and Mo Peterson, old-timers who knew the history of the region going back before the turn of the century. Using a variety of boats and with support from the Super Cub, we oversaw the transition from the small run-down facility to a new construction that would later also become famous as the Wood River Lodge. Forty years later, Alaska senator Ted Stevens and some of our former lodge guests and friends died not far away in the Muklung Hills territory when they were guests at this lodge. I became very familiar with that site when flying wingman with Bob in 5966Xray, his silver Helio Courier. It made me feel like Saint-Exupéry over the Andes. He was my dad's age and had flown bombers over Germany after my dad had flown a glider over the Normandy beaches with the 101st Airborne Screaming Eagles.

At the end of a gloriously calm bluebird day or at the end of a scare-you-to-death day of low ceilings and evil turbulence in that tough country, I returned to fresh baked bread and my sun-browned, sweet-smelling little wife. She bid me come in and I did. Which red-blooded fellow in a plaid wool shirt could resist that come-hither smile? Flying in the bush has many unrecognized rewards. What better description of bliss than a fine woman waiting for you in a little log cabin beside a river full of trout?

There were flights at dawn and dusk over country as lovely as God ever made. We saw 140 shades of green that would have exhausted the vocabulary of an Oxford English professor. The summer's lushness surrendered to the burning reds and golds, oranges and crimsons of early fall that made the blueberry hills look like they were afire. Juxtaposed against the intense frenzy in any salmon stream, the scented alpine meadows were so peaceful behind my dozing head that the rest of the world was made to disappear entirely. Queen Elizabeth at her Sandringham Estate was never given a more colorful gift from the banks of the Babingley River than the blushing bouquet I gathered for my girl back at camp. There were glassy landings on water so clear and deep that it seemed you could see right through to the other side of the world. In all that summer lushness, it was difficult to imagine it surrendering to the white mantle that would lock the land tight, hard, and dark for the rest of the winter.

In the fall we flew to Farewell Lake on the west side of the Alaska Range, and Diane worked with Gayle Curtis at the lodge while I flew hunting clients for moose and grizzly, caribou and sheep. Diane also cooked in a spike camp where the classic white-walled tents set up beside the river left another image indelibly imprinted on both of us. One of the spike camps was next to the long-abandoned Rohn River Roadhouse, although it would be years before the Iditarod Trail Sled Dog Race from Anchorage to Nome was initiated. The old gold-rush roadhouse that we got to know became a strategic checkpoint on that grueling eleven-hundred-mile route years later. When we first saw it, there were dog harnesses hanging on the walls, a few old books, and wool long johns with all manner of other paraphernalia from the miners, trappers, and mushers of a bygone generation. The carefully hewn log cabin was an untouched time-warp museum. Thrown a few feet from the cabin door and overgrown with weeds was a treasure trove of crystals and gold ore samples. We could imagine the prospectors returning to camp with samples of what they had seen, showing them to their mates, logging the finds on their maps, and tossing the samples in a pile; a few of those pieces sit today on a windowsill at the lodge. Bob landed the Courier on huge low-pressure tires on those gravel bars with cabbage-patch surfaces. The leading edges of the slatted wings banged in and out with his full stall landings in such a shocking way that most of the hunters emerged from the swirling dust storm sputtering and pale, eyes bulging as if they had just survived a tornado.

During the first twenty years of our occupancy in China Poot Bay, we had neither the money nor the inclination to own a floatplane and so chartered our diverse flying needs with local air taxi operators. I learned a great deal from local pilots about flying the wild coast and give those pilots credit for the examples they set that helped me fly the intervening twenty years without an accident.

Many Alaskan pilots consider the Piper PA-12 to be the near-perfect bush plane, combining as it does high performance, simple maintenance, the ability to pack large loads, and fuel efficiency. Ours has been completely restored and looks like new. Built in 1946 as a flying ambulance during World War II, it has just enough room behind the pilot for a soldier on a stretcher. That little red airplane became a member of the family. Aladdin himself never had it as good with his magic carpet as we do owning this fine little flying machine. My wife could have been justifiably jealous of my love affair with the Piper. The engine and fuselage were rebuilt time and time again, and today it is as bright as a penny and as sweet to behold as a swallow's wing. There's room in back for Diane, the black Lab, and lots of gear. Visibility is great all around, and this little sweetheart purrs like a Swiss watch.

My old friend Lynn Castle died tragically in a plane crash when he was flying grain for one of his horse-guided trips. He started his lodge on the Wood River south of Fairbanks at about the time we started our own business. He had young children, a fledgling business, and was in the prime of his life. Whenever we lost a fellow lodge owner, fishing or hunting guide, bush pilot, or commercial fishing peer, it reminded us of the tenuousness of our own delicate balance with the unforgiving forces and circumstances that whirled around us.

Alaska's accidental death rate was shocking in those days of primitive or incomplete flying conditions, lack of fast-response lifesaving systems like helicopters, primitive or nonexistent communications from bush locations, poor understanding of hypothermia, unsafe fueling procedures, and the like. Each and every time someone died accidentally in the bush around us we felt as if we had lost a piece of ourselves, and there was never any getting used to it.

The names and faces of a score of friends and fellow pilots speak to me from the grave. In almost every case, they were trying to get somewhere in difficult weather and might just as easily have turned around and flown back home. The thought that these fellows might still be with us, enjoying life as I am, leaves an empty place in the pit of my stomach. Flying in the Alaskan wilderness is always a challenge, and while it offers a sense of wonder that few other situations can replicate, it is replete with dangers for the unprepared or impulsive.

My friend Lowell Thomas Jr. was still flying his Helio Courier 6319 Victor at the age of eighty. Having flown across Africa in a Cessna 180 as a younger man with his wife, Tay, at his side, he has seen more adventurous and dangerous flying than most. We had just flown to a high glacier landing on skis on Denali when he told me about his wife's book, *My War with Worry*. She describes what all of us face: in her case, the challenges of piloting a path through life while waiting for a husband to return from some of the more challenging flying on earth in the Alaskan mountains. The collective residue of fear and grief was for some a heavy burden to carry.

Lowell and I had talked about this a little as he dropped me off on the Ruth Glacier where Steve Hackett, a "grand old man of the mountain," taught me much of what he knew about avalanche and crevasse rescues. The glaciers on our home ground in Kachemak Bay were frequented by climbers, skiers, and pilots sightseeing or landing on skis. I wanted to learn Steve's advanced mountain rescue skills to add to my emergency medical technician certification in case I could be of service to anyone in distress in my area. As a pilot with mountaineering skills, I thought I could help in search and rescue operations from the cockpit if I understood the challenges of climbers: glacier weather, ice conditions, the nature of crevasses and extraction from them, and so forth.

The airplane has given me access to many remarkable places that I would have never known otherwise. For a finely tuned high-performance floatplane in reasonable weather conditions, light on fuel and load, the length of the mountain lake where we built our dream retreat is just about right. Right, that is, provided you are a pilot accustomed to difficult Alaskan bush-flying conditions.

Loonsong Lake is a big lake, but only if you are a small fish. While this little lake is known by a few pilots in the region, its often-mirrored surface is rarely creased by floats other than our own. The lake runs lengthways east and west, up and down the valley. It lies so tight against the steep mountain that in winter, avalanches often sweep whole spruce trees and great boulders far across the thick ice. Those same trees offer still another peril for the pilot as they ghost up and down the lake, almost submerged and as speechless as a Loch Ness mystery. Hitting one on taking off or landing could be fatal. One such monster was so huge that I was viscerally afraid of it. Usually, I attach a line and tie it to the shore and eventually it sinks; sometime this takes years. I was unable to lasso this rhinoceros, but it had a single broomstick branch above water, so I decorated it with bicycle reflectors and multicolored surveyor's tape, lest it should try to snag my floats.

With the glacier just above the east end of our narrow valley and the ocean downstream to the west, the winds move predictably up or down the lake propelled by gigantic forces. Experienced pilots can easily take off and land from much smaller lakes than this one, but Loonsong's length becomes marginal and dangerous if there is low light and glassy water conditions. Added to the challenge are the high thresholds at each end: rising hills with tall spruce trees on their tops.

This hidden Shangri-La is no stranger to moderate and severe turbulence, with downdrafts and gusting winds that my pilot friend Jose DeCreeft describes as monstrous doughnuts rolling down the mountainside. If a pilot descends into such a washing-machine tub and then decides that the look and feel of it precludes a landing, the narrow confines of the valley constrict around the airplane as sinister as a boa. The worst of it comes when lifting one wing, dropping the other, and starting the turn. That's when the knuckles on your right hand blanch white with a grip of steel on the controls and your tongue is suddenly too big for your mouth.

This combination of challenging flying conditions can present a situation in which the pilot is better off simply not landing at all and instead canceling the planned visit. This kind of flying would be easy if the choices were as simple as that, but there are many situations in which the choice between going forward versus turning around is not clear.

Some Alaskan pilots like me who have been flying in the state's beautiful but unforgiving mountains for a long time have clear understandings with themselves about this decision-making process. Before heading out to any place, I say to myself, "I don't have to go there; it will be just as easy to return home and go another time." This clarification makes the difference between old pilots and bold pilots. There are old ones and bold ones but not old *and* bold ones. I have a standing deal with myself: I always get a reward for turning back — a Mexican dinner, a special treat! A draft Alaskan Amber, chips and salsa? A tempting thought. It may be that all day I've been looking forward to that evening sauna and jump in the mountain lake; I've anticipated the blissful sleep in my cold bed under our heavy Hudson's Bay three-beaver blankets and rising early to make a fire and drawing mountain lake water from the pitcher pump to make the first cup of coffee. These thoughts set up a keen anticipation for fulfillment and a feeling of "I really want to get there," whereas offering a happy alternative like the Mexican dinner is a trick — an alternate reward that can tip the scales.

Glassy water is one of perhaps a thousand terms or words more or less unique to flying. This one is relative only to floatplane landings. It is a situation in which the lake is perfectly calm, without the least breeze to ripple the surface of the water to give it definition or texture. The sky is perfectly mirrored in the water surface, so you cannot tell where the surface actually is. Landing on this surface looks straightforward enough to the uninitiated, but treachery lies in wait because the surface is not where you think it is. It is easy to be convinced that you do know because of the surrounding shoreline, trees, islands, and rocky points. When, however, the aircraft has lined up in the standard configuration for its final approach to land, the angles of light and reflection deceive the eye. The eye in turn lies to the brain. With proper training and practice, the brain learns to ignore what the eye tells it in this unusual situation. The brain must insist that the eye look only at the dashboard instruments, not out the windshield at the deceptive surface. In its final moments, a glassy-water landing is an instrument landing. For all intents and purposes, the inside of the windshield may as well be painted black with no opportunity to see out, such is the obligatory reliance on the instrument panel.

When I first started flying into this lake, I was justifiably terrified of crashing into the water when landing. Over time, it became clear that landing east toward the glacier is the safest direction when there is no wind, and the surface is so mirrorlike that treachery lies in wait. This choice is critical. I've landed in my share of places where there was a no "go-around" situation. In such conditions, once you "commit" to landing, it is not possible to change your mind halfway down to the landing site. Pilots prefer to avoid such places unless there is no other choice. At Loonsong, if you land to the west, there is a menacing two-hundred-foot ridge crowned in old-growth timber at the edge of the lake. If you use up much of the available distance with your glassy-water approach and flare, you may not be able to abort the landing at the last minute and go around to try for a new approach. Once you are low and slow you won't be able to climb fast enough and soon enough to clear the ridges and big spruces at either end. The classic, usually fatal hammerhead stall can occur when you pull back on the stick to raise the nose of the airplane and then drop a wing to turn in order to avoid hitting the ridge and trees. The stall occurs when the flow of air over the lower, slower wing is no longer laminar, and into the trees you go for the old dirt nap.

It happens often enough to pilots who lose their sense of depth as they let down. The pilot either slows down too much thinking that he or she is about to touch the water and lifts the nose as the aircraft slows and then stalls and nosedives into the water, or thinks that the water is still below and is going too fast when contact with the water is made. Then the plane might skip off the surface unexpectedly and wildly. If the pilot doesn't apply full throttle and can't clear the threshold to go around and leave the area, it often results in a different kind of stall but with the same disastrous result.

To make a successful landing in glassy conditions, you must be able to bring three of the instruments on your dashboard into delicate synchronicity with one another. If you don't, there are dangers and multiple dire consequences. The aircraft has to be slowed down from the one hundred or so miles per hour that you have been flying. You slowly retard the throttle, watching the tachometer all the while until you see about seventeen hundred rpm registered behind the glass. I usually check my left and right magnetos in the redundant electrical system at this time versus at run-up to be easier on prop wear. At this speed, the little four-cylinder powerhouse that could sit in your lap purrs like a milk-fed kitten. You raise the nose slightly and set up a flight path using more or less throttle to take the craft to the threshold of the planned landing spot. At Loonsong, after you cross the forested ridge, the nose must be pushed down to descend as quickly as is needed. This of course increases speed. The pilot then levels out, and when speed slows below sixty miles per hour, pulls up on the flaps lever. The flaps are the movable aft inboard portions of the wings' surface. These movable extensions of the wing's inboard trailing edge increase lift and help lower the speed. We now have one notch of flaps.

If this landing is late in the day, if it is cloudy and on the edge of dark, then depth and other perceptional senses can be easily compromised. When you are "feeling" your way down to the water's surface like this, you absolutely must not try to discern where the water is by looking out the window. You must concentrate on the instruments. In those very brief seconds waiting for the landing, you feel in the seat of your pants that you are on the very verge of being no longer airborne. There are mere seconds available for micro-adjustments or deciding to abort and go around. Lightning-fast reflexes are needed to pull full back on the stick as soon as the water surface is felt, lest you flip over forward. Or, if you are unfortunate enough to ricochet off the surface instantly, you must apply full throttle and pray that you can clear the trees at the end of the lake as you go around. The combination of all these factors will put your judgment to the severest test. Critical is your understanding of and reliance on your instruments. The smaller the lake, the higher the thresholds at either end, the lower the light conditions, the more dangerous.

When, as a pilot, you finally refine this skill to a feather's edge, there will come a time when the transition of the aircraft from weight bearing on the wings to weight bearing on the water will be so seamless that you will have landed and not even know it. Suddenly a beaming smile will wash across your face with the realization that you just participated in a wonderful magic trick. Welcome to mountain-lake flying.

Having heard all the scary stories I needed to hear and having aborted a landing more than once, I was called on to be creative. Ah-ha! The fix-it guy thought he had a possible solution. From time to time when it was glassy, I had seen a loon or loons on the lake. Small as these little specks were, their wakes gave just enough surface disturbance to provide depth perception and allow a safe landing if it was close to them. But what to do when there were no loons or they were on the "wrong" part of the lake when I arrived on the scene? I got several cabbage-size rocks, flew over slowly, and bombed them into the middle of the lake. I thought that the concentric rings and reflective waves from the shore would provide the surface disturbance I needed. Spitting my gum out the window would have done about as much good. The rocks produced fine waves all right, but they quickly blended into the still surface, and by the time I could circle back to land, the lake surface returned to its same treacherous mirror.

Like any challenge, taking a step at a time leads the way to success, and over time I did become confident with glassy landings on the confined space offered at Loonsong. Perhaps just as important, I got better at reading the mountain weather from a distance. From the lodge living room in China Poot Bay, I could look out the window toward the Valley of Four Glaciers and have a pretty good idea what the weather was doing there. Most important of all, when I flew into the confined valley, I increased my willingness to turn tail and go back home if conditions didn't feel quite right.

Northern lakes go through a fascinating transition as they begin to melt or become rotten; they candle. Water is a near-magical substance that can transform itself from solid to liquid and gas. The shapes available to frozen water are infinite; witness the countless variations of snowflakes. Candles are one of the more curious of the solid phenomena. As the days become longer, the increasing amount of light falling on the ice causes a frozen lake to form floating shapes that, seen in cross section, are hexagonal. These candles can be more than twenty inches long and an inch or two in diameter, and standing on end, leaning against one another, they fool the viewer into thinking they constitute a solid surface.

The Ice Is Going Out

The ice is going out,
Floating on an ebbing tide.
Crystalline waters of freshwater,
rigid on the surface, colliding with the rocky shore.
A billion panes of postage stamp ice, pushed from behind,
each one cherished, beloved, unique in nature.

Transparent houses of cards ascend,
pile one atop another until they can stand no more,
the jumbled ridges collapse in tinkling bells of broken glass.
A slight breeze moving seaward hears that elegiac music of singing angels,
and regret is in their voiceless response.

Anyone who considers going out on a frozen lake needs to be clear about the risks. On my unlucky day, the ice thickness in the morning when chopping a hole to get wash water for the sauna was a measured twenty-three inches. That seemed sufficient for a plane much bigger than our own.

It was midday when I returned from a reconnoitering flight farther up the valley. I wanted to figure out the routing of a trail into the high country that I hoped to use in the summer with young people from the International Wilderness Leadership School. I flared for a normal landing, and as the speed slowed, the gear broke through the ice — not a pretty sight. I scrambled out and stood on the ice almost in tears at the pitiful scene before me. The prop was bent, and the fuselage was easing down into the black water where the bottoms of the wings would rest on the ice. This broad weight dispersal prevented it from plummeting to the bottom three hundred feet below.

The cost of the helicopter extraction, not to mention prop replacement and engine and fuselage repairs, is where the story gets ugly and very uncomfortable to talk about. The ice had candled, and the twenty-three-inch pieces — billions of them — were standing on end and leaning contentedly against one another to remain upright. The physics of this strange structure make the ice look strong from above, but every year in Alaska, airplanes and moose, snow machines and musk ox, not to mention hikers and skiers, come to an unpleasant end by venturing out on early spring ice. For a brief period as winter merges into early spring, this grief can be avoided by being on the ice at first light when the candles are fused together. As spring advances they collapse, and a strong wind might blow them to shore where they clink and tinkle against one another in the waves like some ethereal music from another world. Perhaps no sound in the Far North is more magical than when this music is married to the plaintive yodel of the hermit thrush's song.

When the ice collapses and starts down the rivers, it moves first in a tinkling chorus as millions of these ice candles melodically bump into one another in the current. This changes to a more urgent sound as the flow increases. It is music that to a person's ears and sounds like it could have been made only by angels. The people listen, the wind listens, the fish listen — and the sound is recognized by each.

Alaska's coastal mountain ranges have challenged pilots since Russel Merrill flew the wilderness coast north from Willamette to Anchorage in 1925. More than one of my friends has gone to the stony lonesome at the hands of winds from these deceptively lovely mountains. The Kuroshio, or Japan Current, swirling in from the Gulf of Alaska spawns warm, wet weather of gargantuan, hemispheric proportions. Especially during the changing of the seasons in spring and fall, this soggy battlement of leaden air is hurled against the soaring coastal mountains. The thousand-mile curve of the Gulf of Alaska merges into the Alaska Peninsula as it drops south and west to join the thousand-mile arc of the Aleutians. Our little lake lies at this mighty intersection. This is the place where the weather wages a battle royal with the great glaciers of the Harding Ice Field that rim the northwestern edge of the continent. Here, the Gulf of Alaska, the tip of the Kenai mountain range, Cook Inlet, the end of the Alaska Range, and the beginning of the Aleutian Range all come together. Now throw in a few active volcanoes, and these giants combine to create a veritable witches' brew of weather. Why would a sane person want to fly a floatplane into surroundings such as these?

Getting to know this lake and the particulars of these landings occasionally left me terrified. Having a healthy dose of respect that borders on apprehension — even fear — is a good thing. Being terrified is not a constructive emotion in the cockpit, but being a little scared is. Fear provokes the adrenal gland, whose chemical excretions call dendrites and nerve endings to rapt attention. It is a physiological fact that adrenaline speeds reaction time in the lizard brain's world of flight-or-fight response. This is the environment in which it is essential to keep a level head and respond instinctively and immediately. Bringing the intuitive sense into play is, according to several ancient philosophies, more valid than the more typical mental gymnastics. Intuition is one thing; responding to intuition with changed behavior is quite another. When in the cockpit or in other situations I sense that "I just don't like the way this feels," I listen to that voice and just go back home. There will be no stop and go sign, there will be no one else to tell you what to do. The thin veil between the right and the wrong decision is diaphanous indeed.

Even after many years of flying to and from Loonsong, there are occasions when the go/don't-go conditions are difficult to separate. There are times when the valley is so shrouded in low cloud or fog that a visit seems unlikely. There is no harm in flying up the valley a ways, as long as there is a firm commitment to turn around as soon as conditions look too questionable to continue. Time and again, the heavy, cold air coming down from the glacier slides under or pushes back the moist coastal air and creates clear air over the lake when it is crummy up or down valley. There are other times when the flight to the lake looks possible, only to find it's necessary to turn back just before landing. The surface of the lake is often so ripped and rent by descending whirlwinds we call "cats paws" that one wouldn't think of landing. There are also many times when the go/don't-go decision is a teeter-totter of indecision, and I let the totter teeter in favor of caution.

One year, Morgan and I had been looking forward to a flying camping trip across Cook Inlet for most of the summer, and a fine day finally arrived with more good weather in the forecast. We have shared many great adventures with eight-oh mike, and this promised to be a sweet one. The opportunity to sleep on the ground in a wild place in a tent qualifies as a perfect getaway. The floats lifted off of Beluga Lake in Homer, and we headed west early one the morning to take good advantage of the window of opportunity before us.

Sixty-eight miles away, Augustine Volcano was leaving a faint vapor trail that stretched away to the southwest. Chernabura, as the Russians called it, rose as symmetrical as an inverted funnel out of an unusually flat ocean. Our destination, however, was a favorite camping spot fifty miles beyond it, from which we planned to hunt moose a month later. Supplying our family and friends with that delicious red meat is always a highlight of our year.

After almost an hour of flying at three thousand feet, we approached the halfway point. The volcano looked so attractive and Cook Inlet so calm that I decided to land at its flank with the consideration *in mind* of climbing it. A quick glimpse at the dashboard tide book revealed that the tide cycle was small in vertical change and that it would be going down for a few more hours. We considered our landing spot, skidded to a stop, and taxied slowly and carefully up to the shore. Boulders the size of refrigerators were scattered widely above and below the surface of the water; the spaces between were a flat bottom of sandy volcanic debris. At forty-two hundred feet, the volcano is not an easy climb, especially since its flanks are thickly interwoven with alders. Anyone who has climbed on steep slopes through alders says prayers every night that they will never have to repeat it. They can present as formidable an obstacle as the thickest African jungle.

The bigger issue in front of us was leaving the airplane for the length of time required for the ascent and descent. When mountaineer Galen Rowell and I had climbed together, he told me that he considered an ascent rate of one thousand vertical feet per hour as respectable. Even if we could breeze through the alders, it seemed we were faced with leaving the airplane unattended for at least eight hours. Not a good idea at all. I felt in my bones, however, that the planets were lined up in our favor, and that this was one of those rare days when the typical day breeze would not blow. We did not have an anchor for the plane, but there was a scattering of perfect anchors — those ten-thousand-pound boulders. Quick as a wink we tied the plane in such a way that if the wind did come up, it would be safe. This was a big risk and an optimistically bold assumption. Off we went with backpacks of food, water, and rain gear. Against all odds we found a chute where the melting snows had carried down ten thousand dump-truck loads of pumice in a fairly straight and compacted path. Up it we went, unobstructed by the menacing interwoven tangles of alder thickets on either side. After a few hours, the volcano's slope pitched up and up until the angle of repose of the material under our feet was such that we were taking one step up and sliding two backward.

It became so steep and crumbly that at times we were on all fours. Our eyes were much troubled by the pumice dust, which was as irritating as powdered glass. Each boot step puffed up a small cloud of the silica. Although coughing and on the verge of quitting, we encouraged one another and pressed on. Our spirits lifted when at last the summit was in sight. The thought of what we might see looking down into the caldera excited us, and we plodded on.

At the top, we carefully chose our steps in the broken lava and teetered along the lip, which dropped away into the gaping maw that was half-hidden in steam. Beside one of the black basaltic cliffs, we encountered a fumarole whose orifice was emitting a blasting jet of superheated air. It gave out a hideous sound that was something between a moan, a growl, and a hiss. Timorously, we edged closer with no small amount of apprehension for the danger it posed. It was no bigger than my waist, and around its mouth were beautiful yellow crystals the size of a small hand. We guessed that the substance was sulfur.

"This could be the exhaled breath of the great sleeping dragon, Morgan. This vent connects directly into the very heart of the earth. Let's get the hell out of here before he discovers that we're here and cooks us for his dinner."

It took us a little more than four hours for the ascent and less than three to return. Eight-oh mike was there waiting for us, bobbing in the middle of Cook Inlet like a rubber ducky. We piled in and that night slept in the moose camp on a luxuriant carpet of Labrador tea whose calming breath gave us sweet dreams. During the night, beavers from the nearby lodge repeatedly took exception with the red airplane's presence in what they had claimed as their cove and smacked the water as loudly as a pistol shot. Loons called in the distance; Orion kept watch while his dog, Sirius, slept.

This was what felt like a world-class adventure that was facilitated by the floatplane. It would not, could not, have come about in any other way. There have been scores like this in scattered places all across Alaska, and the patina they have left on my memory and those of friends who have shared them has been something very special.

Although aircraft can provide no end of fascinating experiences and can do many important kinds of work, they can make diverse kinds of social contributions too. I've had the good fortune to put eight-oh mike to work in conservation, "the wise use of resources." The *Steinway* and her pilot were able to play a pivotal role in saving twenty-three thousand acres of the Kachemak Bay State Park from clear-cut industrial logging. Those ill-fated lands surrounded the Wilderness Lodge and the mountain camp, Loonsong. It happened this way.

It would take a sturdy mountaineering backpack to carry all the descriptive adjectives needed to describe the magnificence of the Kachemak Bay State Park. It was the first state park created when the state was still an emergent entity. Its future was considered so bright that acres of land adjacent to it were designated as the Kachemak Bay Wilderness Park.

I had been intimately involved with the early days of the park's creation and had a proprietary interest in the outcome of saving these wildlands. I joined the Kachemak Bay State Park Citizens Advisory Board and later became its chair, as the efforts to save the forests in the park spun out of control.

Having read in national publications about the flying conservationist Michael Stewartt and his creation of the nonprofit group LightHawk — Volunteer Pilots Flying for Conservation in America, I contacted their office to sign up. He too was putting people in his airplane so that they could see the devastation caused by clear-cut logging in the Pacific Northwest and was having some success. My impassioned letters and calls prompted the organization to send their chief pilot all the way from New Mexico to China Poot Bay for a check ride to verify my skill and competency. It happened that on the day I became the first Alaskan to fly for LightHawk, David Brower and his wife, Anne, were celebrating his eightieth birthday with us. When the test pilot, Gene Scarborough, shook my hand in congratulations and got out of the floatplane in front of the lodge, David got in. With the famous "Archdruid" aboard, as John McPhee called him in his book, we flew an aerial ballet I had choreographed that best showed off the biodiversity and scenic beauty of the lands to be logged. David went back to San Francisco, and, using the resources of the Earth Island Institute and his many impressive contacts, we eventually prevailed in saving the park.

As I pulled the floats firmly up on the gravel beach of the tiny island, the magenta-colored river beauty, or dwarf fireweed, danced in a playful mood in the modest breeze. The cool flow had come down the glacier's face and across the lake separating me from its vertical terminus that was splashed with several shades of blue. I had arrived at a far-off lake in a little known place in the Alaskan wilderness that few have ever heard of and fewer still have visited. Because of the distance from the nearest town, it is expensive to get here. Chances are that no human being had ever put a footprint on the tiny beach in front of me. I was at Petrof Glacier Lake on the extreme southwest end of the Kenai Peninsula between the Gulf of Alaska and Lower Cook Inlet.

Petrof Glacier is not easy to find on a map, nor is it easy to access by any means other than the imagination, but having once accessed my secret place with eight-oh mike, I now hold the passport and can visit this remote kingdom at will. Indeed, I often do visit in my dreams. It is a favorite destination on which to concentrate between the times my head hits the pillow and when I am fast asleep.

In my lucky travels to the far corners of the world, like a lepidopterist collecting butterflies, I have caught fleeting glimpses of peacefulness and that most precious of artifacts: silence. I have come to know and treasure the memory of exotic kinds of silence in its endless forms. Whether in the midst of the vast Atacama Desert of the Inca people in the Andes or at Powoovoouliliak on the island of Sivuqaq in the frozen Bering Strait in the land of the Inupiat, the silence is almost always enhanced by the murmur of blowing sand or the abrasion of moving ice or shifting snow. By and large, "achieving real silence is," as described to me beside a fire by a German ambassador when he was a lodge guest, "an expensive matter, involving travel to a faraway place in wild nature without people to find it. There is usually an airplane in this equation."

The simple crunch of gravel at the bow of my floats in this magical place was as sweet to me as the sound of dancing finger cymbals would be to a newly crowned king entering a welcoming citadel. Stepping down from the floats in the midsummer sun I felt like the celebrated son, returning to accept the jewel-studded keys to the city. The fireweed was a gift from a shining Nubian. He presented me with precious silks that were the late summer colors surrounding me. I caught the smell of exotic incense rising from a brass censer — or was it the sticky cottonwood buds in the ripeness of the sun? In place of swaying ostrich plumes fanned by long-haired maidens, mountain goats spotted the slopes above the ice wall where an eagle hung suspended. The incense was that of summer's end, air as clean and crisp as a feather's edge. The Taj Mahal in moonlight had no inlaid lapis lazuli more beautiful than the angled light reflecting from the bergy bits of ancient glacial ice that had grounded on the shore beside my floats.

The admission price for access to such a place is high. Landing on such milky-gray lakes where the bottom is unknown and unseen relies heavily on intuition, and it is risky business. Sandbars and rocks can lie just under the surface. If the wind rises while visiting such a place, it usually comes from the face of the glacier as cool air flows downslope. Making their own weather, such places do not appear on any weather charts or forecasts, and again the role of intuition is pivotal. With few exceptions, floatplanes must take off into the wind. This means flying directly toward the dragon's teeth and climbing steeply, then initiating a turn at the edge of the performance envelope. It is not a pretty picture. Who in their right mind would tickle the tail of a sleeping dragon?

The lake was calm, the breeze died away. I did not expect the dragon to stir. This is an issue that a floatplane pilot must always address when landing on a lake close to its glacier and planning to spend a while on the shore. Seafaring men like my brother call the Aleutians the "place where the great storms are born." Glaciers are nothing if not also creators of great weather. The slightest change of barometric pressure or shift of weather pattern can raise them from dreamless sleep to violent rage. Woe betide the floatplane pilot who ventures too close to the recumbent monster without due respect and a strategy for a rapid exit. Living for forty years at the mouth of the Valley of Four Glaciers in China Poot Bay and inhaling the exhaled breath of these cold-blooded giants, I have come to know their many moods and faces and have at least some dim sense about when it is marginally safe to approach them.

Silence is like wine; there are chardonnays and fumé blancs, zinfandels, cabernets, and ports. There is a special language in use and an air of romance surrounding wine in its many forms. It is the happy companion to candles and a fine meal at a favorite restaurant. Silence, like wine, partners with the sacred on altars, in festivities and celebrations. Silence too figures in human imagination and creativity at every level, and its facets are limitless.

Paint River Falls. Photo © Boyd Norton.

CHAPTER 9

Beneath Cold Waters

"The Sea, once it casts its spell, holds one
in its net of wonder forever."

—Jacques Cousteau

A tiny trickle of ice water has slipped past the face and neck seals of my scuba dry suit. The unwanted guest was worming its way down my neck and began to spread like a fan across my chest. A quick readjustment to the rubber seal stopped it, but not before a rolling shiver crept from my nose to my toes. Standing up in the waist-deep water, I had just taken my first look into the mouth of the Paint River Falls. As far as I knew, no living person had ever seen what I had just observed.

The indigenous people of this area once called it the Achchek. It is a little-known river mouth in Kamishak Bay, which is on the west side of Lower Cook Inlet, Alaska, and about 150 miles south-southwest of Anchorage. The Paint River runs crystal clear and drains much of the McNeil River Brown Bear Sanctuary and Refuge, an area that encompasses the world's largest population of coastal brown grizzly bears.

We were about twenty-five miles inshore of St. Augustine, an active volcanic island, as I wiped the ice water from my face. My diving partner, Jim Sharp, and I had gone to considerable difficulty to arrive at this remote place, and now we were preparing for a dive to investigate a deep, narrow, and mysterious channel that has been carved by the river into ancient bedrock. Being in this wild place during the salmon run means by definition that the area is thick with the great carnivores that pace the shoreline, hungry after a long winter hibernation. The shaggy beasts are interested in only one thing — salmon, and they are swimming all around me.

Alaskan diving offers the rare opportunity to visit that part of the forty-ninth state that is the biggest, strangest, strongest, least known, and the most unforgiving of its many dominions. Few venture under water in the Far North; those who do visit only briefly. There are literally hundreds of thousands of square miles of fascinating underwater terrain that have never been seen by human eyes.

Extreme diving has many similarities to being in space, but there is so much tactile involvement under water that it is a sensory immersion of the first order. In allowing us to breathe under water with compressed air, science and technology have demystified this realm that has remained throughout human evolution a place of mystery, the place of gods and spirits, both good and evil. While modern men and women might interact with this realm in reasoned and practical ways, be they scientific or recreational, it is worth considering how our psyches deal with this activity.

Poetry in its precise science helps us measure our own existence, as expressed in this poem by our lodge guest Barbara Racusen of Shelburne, Vermont, when she speaks of this primal memory.

I do not remember learning to swim, but, I have some sense of
　　place, where,
Before knowing any fear, I first trembled and kicked,
In the warm and salty umbral amnion as undulant as kelp at
　　flood tide,
And mother reached for weeds in July flower beds.
Nor do I quite recall, what was told, retold,
That father, set his sun warmed toddler, belly down, in ocean's
　　glistening swell,
like releasing a too small fish, and tasting salt, I quivered and
　　I swam.

My old friend, associate, and fellow board member Sylvia Earle and I share a glorious sense of completeness by being under water. While most are completely out of their own element in a cold-water diving suit, some of us are actually thrilled to be in that environment. Perhaps the kind of surrender involved here is because of the amount of salt water in our veins. Perhaps the lizard brain remembers being suspended, upside down, in the amniotic fluid before our own birth.

Being comfortably under water resonates with what Rumi says about death: "I have felt nothing, ever, like the wild wonder of that moment" — it is never any farther away than an equipment malfunction or accident. Diving in very cold waters in the Far North provokes a sense of wild wonder such as few other experiences can. I have nothing like Sylvia's vast level of experience, but we share a commitment to helping people understand the importance of taking care of *la mer*, our mother, the ocean. We share the passion to reach for those sublime heights that can only be experienced in the depths. *The World Is Blue* is Sylvia's most recent publication, and it is a clarion call to the entire world to take care of the oceans. These stories, like her own, are intended to evoke wonder in order to provoke action to protect our earth.

In the early boom days of Alaska tourism development, when flying became something that most people could afford, someone came up with a clever promotional mantra: "Alaska — America's Last Frontier." Did anyone ever consider that the real and literal frontier was under water? Alaska's coast is so long and bifurcated that it is measured at over fifty thousand miles in length — more than twice the circumference of the world. Probably well less than a fraction of 1 percent has been "discovered" or visited by a human being.

Of the 70 percent of the earth that is covered by water, more than 80 percent of that is over a mile deep! Those percentages in Alaska would be skewed because so much of the Aleutians, for example, drops directly from the shore to mind-boggling depths. The Aleutian Islands are taller than the Himalayan peaks when viewed from the bottom of the ocean.

Getting under the ocean in Alaska in unusual places could easily fill several lifetimes. The oceans are so fascinating that, by comparison, there is little interest in getting under the freshwater, unless you are a salmon ichthyologist. Landing frequently on alpine lakes with our floatplane provokes a curiosity about diving in such clear-water places. It was virtually certain that no one had ever dived in some of the wilderness places I had in mind, and diving in the volcanic caldera lake of Mount Kaguyak on the shoulder of the Alaska Peninsula before it becomes the Aleutian Islands had long intrigued me. This long-extinct volcano had popped its cork in the distant past and left a great circular hole in the top. Apparently, snow and rain had filled it since it doesn't seem likely that a spring could have done so. A fellow pilot, Ken Day, the certified flight instructor who tests me in my biannual flight review, reported that it did indeed contain trout, not to mention hot springs. There was no rent in the side of the circular peak where the water had overflowed and carved a channel down to a river, so the idea that there would be fish in it seemed preposterous. In order to see it with my own eyes, I planned a camping trip that would last a few days.

In geological terms, the dormant volcano is part of the grand volcanism of the Katmai region whose Novarupta event of 1912 put so much ash and aerosols into the upper atmosphere that it lowered the earth's temperature by two degrees for three years. It was the largest volcanic eruption in one hundred years. On arrival, it was necessary to fly a course spiraling down into the volcano's round mouth and then land on glassy water in tight quarters. And sure enough, there were both hot springs and trout. How the fish ever got there I couldn't even imagine. Meanwhile, it certainly looked like an interesting dive site. Trouble was, I couldn't fly in with a dive partner plus both sets of diving equipment and camping gear. Landing in glassy-water conditions with a heavy load in a place like that was not a good idea. There was an impossibly high threshold and a steep turn if a go-around was needed, and that was clearly a deal breaker. With no possibility of wind to assist takeoff in a small space at two thousand feet above sea level with its lesser-density altitude was another disqualifying problem. Perhaps someone will dive there someday with helicopter support — a venture that will cost many thousands of dollars for those with plenty of money burning a hole in their pocket. I wish them luck.

Back in the Paint River with the bears and the salmon, I said, "Good grief, Jim, this is gonna be excellent, most excellent! Just look at the clarity of that water! You can see from here to Sunday through it." I scooped a handful in my insulated glove. "And boy, does it taste good."

Tasting the water I am diving in gives me a sense of its salinity and bonds me to it. Even before launching over the side of a boat, splashing a face full steadies the mind. This is the right time to draw calmness up from your flippers and let the body have a moment to readjust to what is coming. It has been hundreds of millions of years since our amphibian ancestors were comfortable under water, and failure to acclimate slowly to very cold water causes the panicky brain to demand more compressed air than necessary. Having come all this way to the Paint River, who wants to abbreviate the amount of time under water in a remote place because of rapid versus calm breathing?

As I considered the purity of the water, I recalled my reluctance to taste the water in Micronesia, where we dove at a popular dive site that was badly polluted. What a refreshing and invigorating thing it was to dive in these pristine Alaskan waters that drain a vast watershed without a cabin or trail, without a single indication of human presence. If my olfactory senses retained the sensitivity of the salmon's, I might have been able to describe the differences in the taste and smell of the Boro River in the Okavango Delta in Botswana versus the Zambezi in Zambia and compare each to the Paint. I had been under water in both, where dangers were similar but temperatures slightly different. The pink salmon swirling around our knees here had indeed found their way back to this little gorge across their thousands of miles of migratory routing without a passport and based entirely on their sense of smell. The scientists tell us that they were imprinted to almost the exact spot where the female fanned out a redd in the gravel with her tail, and where the male squirted his milky sperm.

The Paint River Falls and the dramatic gorge they had cut into bedrock first entered my sight from the right seat of Chip Mills's

Cessna 185 when we flew from China Poot Bay to Battle Lake in 1976. Subsequently, I had studied it scores of times with other pilots and finally in our own Piper PA-12 on floats. On that first flight it was my opportunity to introduce Chip, his wife, Beau, and Diane to Ben White and his wife, Whitey, at their camp at the outlet of Battle Lake on the shoulder of the Alaska Peninsula. Ben was a living icon for the bush pilot and wilderness man. In Japan, he would have been formally recognized as a Living National Treasure. His shy humility covered the fact that he had done and seen things in Alaska that are the stuff of legend.

As luck would have it, the day of our inlet crossing dawned bright and clear. We carefully studied our route on a topographical map as we preflighted the aircraft. We noted the falls marked on the maps where the Paint River's huge watershed fell into the ocean. When we got to the falls an hour after takeoff, we were each spellbound by what we saw from above. Although neither of us was good enough at hydrological measurements to take a guess at the cubic feet per second hurtling over the falls, we knew that it would be an impressive number. Notwithstanding any numeral, it was simply a magnificent sight. This torrent of water seemed to go over the edge in the shape of a cylinder. The flow was fat, as heavy and round as a fireplug. Compared to other falls like Niagara or Victoria, it was like looking at a bulldog versus a greyhound. Of course we couldn't hear anything over the roar of the Continental engine, but the look of the falls told you that on the ground its roar could be heard for miles. Who knew if there might be a Tarzan movie-like cave behind the waterfall?

A lifetime of other dives from Africa to Southeast Asia, the Bering Straits, and Polynesia had given me an unusually large perspective from which to assess the uniqueness of this particular setting. I had worked for a few years on the famous world-ranging Lindblad *Explorer* as expedition leader during the Alaskan winters when we had no lodge income. I had been able to create diving opportunities in the most unusual of its far-flung travels. Easter Island, the Aleutians, the Arctic Ocean, and a score of other dive sites had given me a global underwater calculus that put this dive in perspective. On some of the early trips to Antarctica, there had been a rare opportunity to dive among icebergs and penguins while keeping an eye out for leopard seals at Argentina's research station, Esperanza. A sense of nervousness stayed with me as I thought of leopard seals, and I kept my eye peeled for brown bears there at the mouth of the Paint.

While we weren't exactly rubbing shoulders with bears, there were dinner plate–size tracks all over the sandbar where we got into our dive suits. From the sandbar, there was a well-worn bear trail leading up to another trail along the edge of the underwater canyon. For centuries, bears had been walking in the same footprints that left deep indents in the tundra. Fuchsia-colored pixie eyes on their shamrock-green tennis-ball bases were scattered here and there among the tracks, and the fat campanula lanterns hung their showy powder blues over the canyon's void for all the world to envy. Yellow cinquefoil plants sunned themselves on rocky ledges below, sharing the space with a plant nicknamed "Eskimo carrot," a lovely little umbelifera.

We had hiked that trail above the pool early in the morning, trying to assess from above what we might encounter in the mysterious cleft below. Schools of salmon nosed into the currents that shimmered with emerald and aquamarine shadows. We could see from the several directions they pointed that there was a complexity of currents generated by the hydraulic of the huge waterfall just up-current and out of sight around the bend.

Since we had seen an abundance of fresh bear scat and half-eaten salmon carcasses, it was clear that there were plenty of the big bears in the immediate vicinity. This salmon-choked gorge was, after all, one of the magnets that drew the giant bears from hundreds of square miles. As we began our dive, we realized there was every chance that a hungry brown bear might have peered down from above wondering how we might taste. At least there weren't any of the Okavango crocs that still terrify my nightmares, no Sala y Gómez requiem sharks or schools of circling Sea of Cortez hammerheads to fret about. A long time ago I had learned to focus my diving attention on more insidious and less visible dangers than the charismatic megafauna.

Having Jim with me was reassuring and certainly safer than diving here alone. That would have been simpler in logistical terms but a serious no-no in diving protocol. In multiple incidents, I had broken the golden rule and nearly made a widow of my wife in the most unsuspecting of circumstances.

On an unusual dive like this, I wished for friends whom I hold in high regard: Carole Baldwin and Tom Barron. We met in the Museum of Natural History at the Smithsonian when Carole premiered her new film for the National Board, *Galapagos: A Submersible Adventure*. In the film she emerged as a perfect role model for any young girl interested in science. Tom writes terrific adventure books for young people ages eight to thirteen. Of special interest was one about a young girl like Carole who has imaginary adventures beneath the sea in a submersible. When Carole told me she had not heard of the book, Tom kindly sent her a copy. When anyone knows that a great adventure is at hand, there is a wish that others might share it.

Sinking below the Boston Whaler, Jim and I turned our backs on the last sign of humans and entered a realm where titanic forces over the course of tens of thousands of years had shaped forms unlike anything either of us had ever seen — unlike what wind-driven snow and ice could ever shape above. Icebergs show the curvaceous shapes of water, sun, and wind on their flanks. Snow in its many moving forms can take on exquisite forms, but here were flowing sinusoidal lines that spoke of monstrous forces over vast periods of time working with endless patience on solid rock.

The light in which we hung suspended called to mind astronauts, unencumbered by gravity, floating free in a third dimension. We felt as weightless as the light itself, made so by the air trapped in our pressurized spaceship-like dry suits. We descended with a twist of the shoulder-mounted bleeder valve that released air from the suit and stopped the descent with a light tap on the equalization valve. With that tap, a hiss of compressed air entered at the left nipple to add air into the suit to compensate for the increased compression of the descent. It felt like magic to be suspended thus in gin-clear water.

My friend Dr. Ian McCallum, psychiatrist and medical doctor, Jungian, and Springbok rugby player, in his powerful book *Ecological Intelligence*, makes bold to talk about poetry in situations like ours. Whenever I readjust my buoyancy compensation so that I neither plummet to the bottom nor rocket to the surface, I consider his comments about how poetry helps us achieve balance through the guidance of carefully chosen and scripted words. The poet, like the diver, must have a keen sense of balance in order to be successful in his or her practice.

Our downward glide slowed until we hung suspended just above the bottom. The surface above sparkled like a burnished mirror. Could that brown shape at the edge of the canyon be a bear peering over the edge and wondering what kind of large fish those were swimming below? Was that a bit of Pavlov's drool on its lips? Was he considering us for his dinner?

Pale spectral colors danced around us as if shot through a stained glass. The canyon called us onward, its bottom and sides undulating in and out, up and down. One might as well have been a tiny bug within the thirty-three-foot transit of a French horn. The groins and arches, parapets and flying buttresses were reminiscent of the medieval castles of the Knights Templar. There were chambers and ballrooms, unexplained convexities and polished concavities. At each curve in the path we paused for long moments to take in the magnificence that was around us and to consider what was ahead. We were simply spellbound.

There were no industrial sites or upstream community runoff to muddy or pollute the water, no pop cans or tires littering the bottom, no fishing lures snagged between rounded boulders. There was not a single clue that this scene had changed one iota in the last ten thousand years. Indeed, in the surrounding two hundred square miles of the Paint River watershed, there was not a single road, cabin, or even a human footpath. It was as holy, pure, and untouched as when God made it. The cathedral metaphor pales in insignificance. Those human-made creations are in the end manifestations of humans' intellect as well as their financial and logistical maneuverings. As often as not, they were created by impoverishing, crippling, or even destroying human life to accomplish an ego-driven goal. Here was the very hand and face of God, at once diligently working and quietly at peace.

The susurrating background hiss that we could hear up-current was that of countless tons of water falling from the cliff eighty or so feet above and into the frothing pool below. I found it strangely unsettling.

As I hung there motionless in the frigid water, a flashing school of bright salmon broke the trance. Gliding past were hundreds upon hundreds of content-looking silvery messengers bringing unspoken greetings. They were so bright that they had probably just arrived from the open sea, perhaps carried on the same tide that had allowed our boat to cross over the barrier sandbar and into the mouth of the river canyon. Their slow passage was as peaceful a scene as I have ever experienced. If they carried news, they were unhurried about delivering it. They seemed to have no place to go. I was of as little interest to them as the round boulders on the riverbed below us.

It was difficult to focus my attention in that dreamlike state of suspension. Everything seemed so simple, but I kept trying to compare what I was seeing to other observations, and nothing seemed to come close to this magnificence.

Since time immemorial, the confines of the canyon had been lovingly caressed and savagely abraded by the slow adagios and speeding allegrettos of countless billions of individual grains of sand. Grinding their way at last to the ocean under the baton of the Great Conductor, the pebbles and boulders each sang with their own distinctive voices. Large and small, each had moved down from the mountains and val-leys miles away and far above as if to prove that some omnipotent force wanted the mountain laid flat. Far away, rain and melting snow found its way into the tiniest crevasse on the highest pinnacle. There it penetrated, froze, and broke the stone. Cracking and cracking again, avalanching and falling, the pieces were spread into alluvial fans to rest for perhaps a thousand years. Then they might have been pushed downslope by water, snow, and ice once more. Eventually, a stone from the tallest peak might have found itself in this very pool on the way to the ocean.

It was that lithic-graduated assortment, however, that won the prize for unexpected discoveries on this dive. When we had stood on the bear trail at the edge of the cliff, we had no clue what the bottom would look like. Now that I was down there, I wondered if perhaps I should be looking out for the ghost of a finned Sisyphus forever rolling boulders back upstream. Because of their great size, some of these boulders had remained in situ for centuries rather than being blown out of the canyon. The hurrying water could not get an effective grip on their sides to move them any farther. Acres of round stones appeared, carefully sorted according to size. Some alleys were littered with medicine balls, corners were crowded with only soccer balls, and narrower slots were filled with marbles.

As we swam slowly toward the waterfall, we encountered embayments and coves. Like the sweeping arches high overhead in a medieval cathedral, there can be grandiosity in simplicity. Here and there was a monster rock, sleeping the age of the mountains, so enormous that each stirred restlessly perhaps only an inch or two once a century when the earth itself trembles beneath the onslaught of the great spring floods and booming waters. Knowing we could never come this way again, we swam very slowly, hovered, moved forward again, just a little. We didn't want this to end.

Floating among the boulders with my imagination in full bloom, I felt like I was on my way to becoming one of the salmon suspended in the current in a slanting submarine shaft of light; I was one of the swirling bubbles in the wiggling curve of my own ascending air. It was a mixed-up crazy world down there. I felt bigger than myself yet small and vulnerable. The dancing beams of chaotic light called my attention to concerns about vertigo, and yet the sensitive balance requirements that I manipulated gave a feeling of being in complete control.

On my right and among some boulders, appearing as if in a dream, I became aware of a single object that seemed to be glowing. The bent rays of light in the undulating halocline were confusing me. The word *vermiculated* describes the squiggly trails left by worms, and it perfectly describes the way the light responded to the sandwich layers of fresh- and salt water. The freshwater was from melting mountain snows, and the saline waters swept in from the wide Pacific upwelling. The two were so different in chemistry and density that, as they moved and commingled, they reflected light like worm tracks. It reminded me of what happens to refracted light above a candle or what you see if you look closely at the exhaust fumes coming from your gas tank when you are filling it.

The weird glow came from a concavity, not from a cave, which we had agreed not to enter. I paused before continuing. There were no overhanging slabs or boulders. A jester, a trompe l'oeil held sway here; what the eye saw was not in sync with what the brain perceived. The otherworldly glow was in an enclosure like the niche in a cathedral wall. Here was the calm of intimacy beside the grandiosity of the larger space. Was I looking at a piece of a crashed airplane or boat wreck or some kind of luminescent highway marker? We had not seen a single trace of anything human related until this anomaly. As I curiously approached the luminescent glow, I saw that it was a large rotting king salmon so large and grotesque that it gave me the heebie-jeebies. I have seen thousands of dead and dying salmon over the years but never anything like this.

I approached with the curiosity of a cat with a snake, slowly, trying to make sense of what I was seeing. I would discover only vacant, blind, staring eyes, bared teeth, and flesh rotting from fin and bone. It was corpselike, horrible — not at all like any of the dead fish I had seen above the water. Somehow, the milky and clouded putrefaction of the fish absorbed or reflected light like nothing I had ever seen under or above water. It was a ghostly reminder of our own brevity and the unrecognized dangers around us.

As birds know the skies, so fish know the sea; so I did what I saw the schools of salmon doing as they hovered motionless for long moments. In the air, I would have been a small bird suspended in front of a nectarous flower. I was trembling a little, but it was not from the cold. An uncertainty was enveloping me as I turned my back on the unseen falls, the attraction that had lured me here in the first place. Something said simply, "Turn back." Just ahead had been the imaginary pot of gold in Tarzan's cave behind the falls, and the end of the rainbow's long quest. Was I going to turn back now when I had come all this way to see what the falls looked like under water? Did it hold enchantment or poison? I had a sixth-sense awareness that there was a force at hand on the verge of pulling me out of neutral buoyancy and pummeling me to death by drowning in the hydraulic circular undercurrent beneath the countless tons of falling water.

I slowly began to realize as I swam against the current pulling me toward the falls that my struggle might be too little too late. Apparently Jim had felt what I felt, had turned as I turned, and together we felt the flow of adrenaline as we tried to flee. I knew that in the cumbersome dry suits with thirty-two pounds of lead around our waists, any struggle would be short. Before we had time to think about it we were being forcefully blown out by the reverse flow. We had been grabbed by a force rushing away from the falls. During my emergency medical technician training years, I had watched a training film about the dangers of a similar kind of sucking hydraulic action in a river spillway. We trainees sat at our desks watching helplessly as a man was pulled into the vortex and drowned in a similar situation.

Breaking the trance and squirting out of the hydraulic pull, I was again surrounded by a flashing school of bright salmon, gliding effortlessly, contentedly.

Released from an unseen force that could have killed me, I looked up to the surface and could see billowing clouds high above. I considered what it would be like to look up from that exact position in the dark on a moonless night with a million stars above. I knew that only the salmon would ever know such a sight. The view would be like what the Anasazi would see deep in their southwest canyons, looking up at the narrow band of stars moving quickly across the slim gap above. The submerged shapes in the gorge were reminiscent of the landforms sculpted by water in Monument Valley in Arizona and Arches National Park in Utah. The caves and ledges Jim and I beheld could have been home to Anasazi cliff dwellings at Havasupai. Sweeping my hand now through my exhaled bubbles, I was a god, creating a Milky Way of flung stars.

That evening there was a celebration, a big beach fire. Sparks lofted high into the cool air from the red-and-white cedar driftwood I had gathered. The aroma was a heady spice anointing a remarkable day. We watched the moon rise over Chenik Head and Akjemguiga

Cove whose big tide had carried our boat over the gravel bar and into this little-known slot in the bedrock. We savored the sweet taste of accomplishing something special that we might not ever tell anyone about. Smoked salmon, Pilot Bread, and two brewskis put the icing on the cookie. Unless mountaineers like Galen Rowell or Yvon Chouinard write stories, they generally keep the details of their ascents to themselves. It took twenty years for this one to ferment and mature to a point where it could be brought up from the basement, dusted off, uncorked, and a glass of it poured to share with friends.

Nature's bounty—the mussel beds during low tide at the shore of Kachemak Bay Wilderness Lodge. Photo © Boyd Norton.

CHAPTER 10

Looking Back

The miracle and unique complexity of
wild salmon runs is as precious as anything I can think
of. What is good for the wild salmon runs is inevitably
good for us all. By ensuring their continuity
we look towards our own.

—Rika Mouw, Alaskan artist

With their fellows around the world who are the keepers of wild areas, Alaskans and other Americans share the difficult challenge of balancing the need for resource extraction with the values of wilderness. The money from logging, oil, fish, and minerals is needed to fund schools, roads, and the myriad of public needs. Too often in Alaska and elsewhere, decision makers have come to believe that the debate is an either/or struggle. In our global community and information age, we can and must learn to understand and appreciate the diverse values of wilderness. As I write this, the World Bank is engaged in a major study verifying that ecotourism has positive impacts on economic systems far beyond the face value of money injected into the grassroots levels of the areas. Investment portfolios structured with an eye toward sustainability have increasingly come to consider the financial value of wilderness.

What I saw before me on that sun-splashed day in 1966 from the high vantage point above the bay could well have been the world of monks and poets, mystics and dreamers, were it not for the bush pilots and trappers, homesteaders and commercial fishermen who were there. It was clear in any case that the stupefying power of the scene was unaffected by anything more than a tiny pinprick of human activity. This wild-storm and tide-driven panorama before me had remained little changed since the vast ice sheets melted, the sea rose, and glaciers retreated to reveal land where the First People could settle so many thousands of years ago. We would, however, see in a single generation changes to this panorama that almost defy the imagination. If I had any sense of the mind-boggling changes that our fellow bipeds would bring to such a remote and little-known wonderland, I might have turned tail and headed for some even more remote place in Africa or the Amazon. Then again, having acquired some intimate knowledge of those places, I sadly know that they too have been changed irreversibly and forever.

There are of course reasons to celebrate when wildlands are saved or protected from the crucible of what some call "progress." On my office wall is a photo of the Alaska Department of Natural Resources Commissioner John Shively, Governor Walter Hickel, Mead Treadwell, Diane, and me on the deck at Land's End overlooking Kachemak Bay with its mountains and glaciers. We were celebrating the successful legislative appropriation of $23 million to save the forest at the heart of the Kachemak Bay State Park from industrial clear-cut logging. Thank you Michael Stewartt for the inspiration, and thanks to a bush plane for its role in saving a priceless treasure.

The sad fact is that those of us who care so deeply about saving wildlands and an unpolluted ocean for our grandchildren can never cease our efforts to protect them. Once these values and the integrity of these ecosystems are lost, no force can put them back together again. This is a sobering reality, and it makes all the more joyous the success we have achieved.

Nature Needs Half is a cry raised by The WILD Foundation. At wild.org, my readers can educate themselves about the simple common sense and the clear necessity of saving half of the world's lands and waters in their natural state in order to sustain the burgeoning human population. The lungs of the world, the Amazon Basin in South America and the Congo Basin in Africa, contribute a large percentage of the oxygen that we breathe; the clean sea waters give us every second breath. Our innate ecological intelligence must catch up with our environmental awareness as we go forward.

Edward Abbey in *Desert Solitaire* said it best when he told us to get out there and enjoy the wild places, raft the rivers, hike the mountains, and listen to the night sounds by the campfire.

When Grant Sims, author of *Leaving Alaska*, paddled and hiked with us in China Poot Bay, we talked at length about how we kept our heads on straight in the face of some of the abuses of nature we were forced to watch. I was younger then, and a good deal angrier — forgiveness and wishing the perpetrators the same health and happiness that I wished for my children and neighbors was a big reach for me. Like so many other friends, he too, much as he loved our wild northern land, just couldn't love the land enough to overcome the difficulties with the people who abused it so egregiously.

Our lives in China Poot Bay shot us off on trajectories that we could never have imagined, including the connection we made between set netting on the west side of Cook Inlet and remotest Africa. We fished Hank Kroll's set-net sites in Tuxedni Bay for the Snug Harbor Cannery in 1978 and high lined the inlet. Having hit the mother lode, we paid off the lodge mortgage, bought the adjoining ten acres of land, and bought a sawmill, a John Deere crawler, thousands of dollars of the best Milwaukee tools and chain saws, and had enough money left over to go to Africa with five- and seven-year-old children. We had a storybook adventure living in a remote mud-walled camp where game rangers Chris and Charlotte McBride had just discovered *The White Lions of Timbavati*.

That first love of Africa blossomed into a lifetime of return trips and involvement with African private aviation. Cofounding a nonprofit aviators group with my sister/friend Nora Kreher, we were able to organize The Bateleurs, "Volunteer Pilots Flying for the Environment in Africa." While sitting on the board of The WILD Foundation, an international wilderness organization whose origin was on the banks of the Black Umfolozi in southern Africa, I lobbied members to select Anchorage as the host city for the 8th World Wilderness Congress. When held in Anchorage in September of 2005, the congress was attended by twelve hundred delegates who came from sixty nations to confer on global wilderness issues. This organization and its niche-filling works put it in the place of being the voice for wilderness, worldwide. From China Poot Bay to Africa is quite a stretch; as a matter of fact, it lies at approximately the opposite point on the globe — Alaska's antipode!

One of our clients, who became a good friend and who was a Swiss-born senior partner at Brown Brothers Harriman in New York, nominated me to the National Board of the Smithsonian Institution. He saw to my election for a three-year term, and I was asked by the board to serve a second three-year term. Not only was I the first Alaskan among that group, but I hailed from China Poot Bay, where we didn't own television sets and didn't even get regular newspaper deliveries. Surely I was the most unlikely member of the board in the 150-year history of the Smithsonian. Our experiences in the capital city were something truly remarkable.

The seating around the grand table in the Smithsonian Castle put me between Secretary of Commerce Norman Mineta, Senator Alan Simpson, and other of their peers. After an address by Ruth Bader Ginsburg, I had the chance to introduce Diane to Sandra Day O'Connor.

"Hello, Diane, I hear you are from Homer, Alaska — I've been in the Salty Dawg."

Imagine sitting on a stool at the bar in the Dawg and asking the name of the person next to you, only to discover a Supreme Court justice! I sponsored the first-ever trip to Alaska for the board and the secretary, and it was a great success as they toured the North Slope, visited Prince William Sound and an archaeological dig in Kodiak, attended a dinner hosted by Wally Hickel at the Anchorage Museum, and then on to the Pratt Museum and China Poot Bay.

Morgan and Shannon were homeschooled at the kitchen table as the lodge grew. Sunday *New York Times* and *Connoisseur Magazine* articles triggered dozens of other articles in many languages. Clients came from around the world, and the children's educations expanded as those guests from Germany and Japan invited us into their homes. Our guests added immeasurably to our appreciation for different cultures around the world. European royalty and diplomats; conductors; artists; famous TV, movie, and radio personalities; prize-winning authors, including a Poet Laureate; a member of the president's cabinet, complete with Secret Service escort; and leaders of some of the world's best-known corporations and enterprises all added something special to our lives and to the restored log cabins.

Shannon graduated from Clemson University and worked for Paul Brainerd of the Brainerd Foundation during the first days of its organization. She worked for Bill Gates's father at the Gates Foundation before returning to Homer. She and her husband, Chris, a civil engineer, built a home next to our own in Homer.

Morgan worked for the Kachemak Bay State Park for years as a member of the teams that built the park's trails, and he spent two winters with Mother Teresa's Sisters of Charity in Calcutta and in the leprosy wards at Prem Dan. He has also worked with Homer's own Jodi Miller and her Peace Feather initiative in children's health care in Peru. Any parent's proud comments about his or her children are bound to be suspect, but these two children have inherited from their mother a boundless amount of practical compassion that is a lesson to me every day.

Morgan and Shannon are stepping up to the plate to help us

decide what the next forty years of our business in China Poot Bay will look like. We all agree that the time has come for us to make a greater and more strategic social contribution to the bay area, the state, and the world. At the moment, just what this will look like is uncertain, but we are having a great time working on it.

The yin and the yang, the dramatic opposites of the tides and the seasons, the brilliant summer days and long winter nights, somehow in their bold opposites struck a chord that resonates in the hearts of those soft enough to hear it and strong enough to feel it, to become one with it. Just as some musical forms resonate with some people and not with others, what we heard and felt did not work for everyone.

When we consider a spectacularly productive ecosystem such as Kachemak Bay, it serves us well to remember that there has been no human intervention to make it so. No one owns the land or maintains the fences. There are no barns, tractors, plows, or fertilizers. There are no government subsidies or expensive administrative processes to create this fabulous wealth. This production and distribution system has been in full swing for thousands of years, and unless it is changed by adverse human activity, there is no reason to think that it will not continue while the ages roll. We must do only one thing, one simple thing: do not kill the goose that lays the golden eggs. During the administration of Alaskan bush pilot–homesteader Jay Hammond, Bristol Bay, the mother of waters into which all the salmon streams of this region flow, was put off-limits to oil development. America's insatiable appetite for more and more energy and the state's eagerness to supply that demand has put this moratorium at grave risk. I became involved in a lawsuit to save Kachemak Bay from oil development, and later the governor would appoint me to a blue ribbon commission to review how the state goes about leasing offshore tracts. We would watch over the next three decades as protections were eroded or dissolved, as closed areas were opened, and finally in the midst of the emasculation of key safeguards, *Exxon Valdez* spilled more than 11 million gallons of crude in Prince William Sound. Some 1,300 miles of coastline was fouled as wind and tides carried the pollutants to bird colonies and salmon streams, sea otter and sea lion haul-out sites. Twenty years after the spill Exxon still owed $92 million.

"I came for the north but stayed for the west."

In that vein, I came to find a simple life immersed in nature and instead spent much of a lifetime battling and struggling to protect and preserve that which gave me strength, all the while trying to present to people experiences so powerful that their lives would be shaped by those encounters with nature and wilderness.

Time and again the diverse Balm of Gilead, the down-valley breeze of sweet-scented healing cottonwood buds, the pungent smells of the low tide, and woodsmoke from the sauna and smokehouse tripped that most powerful of the senses in the lizard brain and kept my eye on the prize with my nose to the grindstone. If you've got to do this work, then thank God for this office. The breeze blows through it, the rain and snow fall squarely on my shoulders, and with no roof, the church of the blue sky tells me all I need to know about life here and living.

I was a plaintiff in a successful state supreme court lawsuit to reverse the $25 million sale of oil leases in Kachemak Bay and featured in a *Washington Post* full-page ad that called on America to protect Alaska's wildlands. As the chair of the State Park Advisory Board leading up to the sale of twenty-three thousand acres of the park for clear-cut logging, I would call to order the first formal meeting of the Citizens Coalition and travel to Washington, D.C., where we would raise the first $20,000 to hire professional help and eventually turn back the saws. The coalition raised $23 million from the legislature to do it. As a founder of the Center for Alaskan Coastal Studies and a Smithsonian National Board member, I, along with my colleagues, would have ample opportunity to be engaged at the highest levels of decision making, but even these involvements magnified by ten or one hundred were paltry compared with the forces mounted against us and content with nothing less than our defeat and humiliation.

Michael Carey, editor of the *Anchorage Daily News*, wrote an editorial in which he said that "to advertise yourself as a conservationist or environmentalist in Alaska today, is to risk public disgrace." It is a chilling but accurate observation. A near-lethal arrow that was to rip through my heart in the years to come in a great battle I would lose to protect from trophy hunters the world's largest concentration of brown bears near the McNeil River State Game Sanctuary and Refuge.

We can only hope that more and more people will accept what they are being taught through all forms of human communication about the importance of taking care of our mother, the sea, our father, the land, and the animals as our teachers.

As the years fly by, I would like to see myself tottering down some small path, insatiably interested and keeping old age at bay with my black Lab, whose name is Curiosity. It is this great delight in the simplest things of nature, be they found beside a remote Alaskan lake or in a city park that will keep us forever young.

We came down from the air, from the stars, from a place of mystery, from the breath of God if you will, or from some cosmic force. Leaving by small boat to cross an unknown body of water, leaving any place of safety and comfort in search of something else, is what people have always done.

Mardy Murie had this to say about Alaska's future when she wrote her classic account of life in the north in 1943. In *Two in the Far North*, she says, "My prayer is that Alaska will not lose the heart-nourishing friendliness of her youth — that her people will always care for one another, her towns remain friendly and not completely ruled by the dollar — and that her great wild places will remain, and wild, and free, where wolf and caribou, wolverine and grizzly bear, and all the arctic blossoms may live on in the delicate balance which supported them long before impetuous man appeared in the North. This is the great gift that Alaska can give to the harassed world."

A few things have changed in Alaska and the world since this was written. Her words express the hope that most, if not all, Alaskans support, but they leave it up to each and all of us to decide how to bring this about.

We have known since Plato that education is the key: people respect what they are taught, they take care of what they respect, and they love what they have been taught to respect.

We all know that we cannot fully love another until we love ourselves. Self-respect and self-love radiate out to others until the world becomes a better place.

And now, after more than four decades of total immersion in this resonant harmonic, I feel that ardor for adventure cooling to some small degree. In tune with the natural cycles of life, my grandchildren remind me that this is the time when one can turn inward the energies previously directed outward, toward the quiet and reflection afforded by the passage of time.

Bibliography

Abbey, Edward. *Desert Solitaire*. New York. McGraw-Hill. 1968.

de Laguna, Frederica. *The Archaeology of Cook Inlet, Alaska*. New York. AMS Press. 1975.

de Laguna, Frederica. *Fog on the Mountain*. Garden City, NY. Doubleday. 1938.

Dinesen, Isak (Karen Blixen). *Out of Africa and Shadows on the Grass*. New York. Random House. 1938.

Earle, Sylvia A. *The World Is Blue: How Our Fate and the Ocean's Are One*. Washington, D.C. National Geographic. 2009.

Finney, Ben R., and Jones, Eric M. *Interstellar Migration and the Human Experience*. Berkeley, California. University of California Press. 1985.

Fitzhugh, William, and Crowell, Aron, editors. *Crossroads of Continents: Cultures of Siberia and Alaska*. Washington, D.C. Smithsonian Institution Press. 1988.

Ford, Corey. *Where the Sea Breaks Its Back: The Epic Story of a Pioneer Naturalist and the Discovery of Alaska*. London. Gollancz. 1967.

Foreman, David. *Rewilding North America: A Vision for Conservation in the 21st Century*. Washington, D.C. Island Press. 2004.

Frohlich, Bruno, Harper, Albert B., & Rolf, Gilberg, editors. *To The Aleutians and Beyond: The Anthropology of William S. Laughlin.* Copenhagen, Denmark. The National Museum of Denmark, Department of Ethnology. 2002.

Gardner, Florence Edwards. *Cyrus Edwards' Stories of Early Days and Others in What Is Now Barren, Hart & Metcalfe Counties.* Louisville, Kentucky. Standard Printing Company. 1940.

Haines, John. *The Stars, the Snow, the Fire: Twenty-Five Years in the Northern Wilderness.* St. Paul, Minnesota. Graywolf Press. 2000.

Hammond, Jay. *Diapering the Devil: How Alaska Helped Staunch Befouling by Mismanaged Oil Wealth: A Lesson for Other Oil Rich Nations.* NOTE: This paper is available online under this title.

Hasselstrom, Linda. *Land Circle: Writings Collected from the Land.* Golden, Colorado. Fulcrum Publishing. 2008.

Hay, John. *The Immortal Wilderness.* New York and London. W. W. Norton and Company. 1987.

Jacobsen, Johan Adrian. *Alaskan Voyage 1881–1883: An Expedition to the Northwest Coast of America.* Translated by Erna Gunther from the German text. Chicago and London. The University of Chicago Press. 1977.

Klein, Janet R. *History of Kachemak Bay: The Country, the Communities.* Homer, Alaska. The Homer Society of Natural History. 1981.

Lord, Nancy. *Green Alaska: Dreams of the Far Coast.* Washington, D.C. Counterpoint. 1999.

Lopez, Barry. *Arctic Dreams: Imagination and Desire in a Northern Landscape.* New York. Charles Scribner's Sons. 1986.

Martin, Vance, and Muir, Andrew, editors. *Wilderness and Human Communities: The Spirit of the 21st Century.* Proceedings from the 7th World Wilderness Congress, Port Elizabeth, South Africa. Golden, Colorado. Fulcrum Press. 2004.

Murie, Margaret E. *Island Between.* Fairbanks, Alaska. The University of Alaska. 1977.

Murie, Margaret E. *Two in the Far North.* New York. Alfred A. Knopf. 1962.

Norton, Boyd. *Alaska: Wilderness Frontier.* New York. T. Y. Crowell. 1977.

Norton, Boyd and Yevtushenko, Yevgeny. *Divided Twins: Alaska and Siberia.* New York. Viking Studio. 1988.

Osgood, Cornelius. *Ethnography of the Taniana.* New Haven. Yale University Press. 1933.

Rowlands, John J. *Cache Lake Country: Life in the North Woods.* New York. Norton. 1959.

Rutzebeck, Hjalmar. *Alaska Man's Luck and Other Works.* Santa Barbara, California. Capra Press. 1988. Compiled and edited by Carl Branson.

Sims, Grant. *Leaving Alaska.* New York. Atlantic Monthly Press. 1994.

Thomas, Tay. *My War with Worry.* Ada, Michigan. Chosen Books. 1977.

Wavell, Archibald Percival, Earl of Wavell, editor. *Other Men's Flowers: An Anthology of Poetry.* New York. Putnam. 1945.

Wilson, E. O. *The Naturalist.* Washington, D.C. Island Press. 1994.

Wohlforth, Charles P. *The Fate of Nature: Rediscovering Our Ability to Rescue the Earth.* New York. St. Martin's Press. 2010.

Wohlforth, Charles P. *The Whale and the Supercomputer: On the Northern Front of Climate Change.* New York. North Point Press. 2004.